Vienna Circle Institute Library

Volume 3

Series-Editor:
Friedrich Stadler
Director, Institut Wiener Kreis
and University of Vienna, Austria

For further volumes:
http://www.springer.com/series/7041

Institut Wiener Kreis

Society for the Advancement of the Scientific World Conception

Editorial Address:
Institut Wiener Kreis
Museumstrasse 5/2/19, A-1070 Wien, Austria
Tel.: +431/5261005 (international)
or 01/5261005 (national)
Fax.: +431/5248859 (international)
or 01/5248859 (national)
email: ivc.zuef@univie.ac.at
homepage: http://www.univie.ac.at/ivc/

The titles published in this series are listed at the end of the volume.

Nicholas Tranquillo
Editor

Dream Consciousness

Allan Hobson's New Approach to the Brain and Its Mind

Springer

Vienna Circle Institute Yearbook

Editor
Nicholas Tranquillo
Harvard Medical School
Salem
New Hampshire
USA

ISSN 1571-3083
ISBN 978-3-319-07295-1 ISBN 978-3-319-07296-8 (eBook)
DOI 10.1007/978-3-319-07296-8
Springer Cham Heidelberg New York Dordrecht London

Library of Congress Control Number: 2014945251

Printed on acid-free paper

Springer is part of Springer Science+Business Media (www.springer.com)

Editorial

Since 2001, the Institute Vienna Circle annually organizes the interdisciplinary Vienna International Summer University – Scientific World Conceptions (VISU/SWC) at the Campus of the University of Vienna. In 2010, the general topic was on "The Sciences of the Conscious Mind," with Uljana Feest (TU Berlin), Owen Flanagan (Duke University), and Michael Pauen (HU Berlin) as the main lecturers. As distinguished guest lecturer, we succeeded to engage J. Allan Hobson (Harvard Medical School) and on the occasion of the tenth anniversary of the Summer University, we decided to publish his renowned *William James Lectures*, commented on by experts in the fields of cognitive science, brain and dream research. This volume is the result of this joint effort, for which I am grateful to Allan Hobson, all the commentators, Nicholas Tranquillo, and Robert Kaller from the Institute Vienna Circle.

Vienna
July 2013

Friedrich Stadler
Professor for History and Philosophy of Science
University of Vienna
Institute Vienna Circle (Head and Director)

Foreword

A century separates Sigmund Freud's and Allan Hobson's interpretations of the putative functions of dreams. This "Vienna Circle" volume closes the cycle.

This book consists of three lectures by Allan Hobson and commentaries written by outstanding scholars whose expertise covers a wide range of scientific disciplines including philosophy. It illustrates impressively the extent to which scientific enquiry from a third person perspective can contribute to the understanding of phenomena accessible only from the first person perspective such as dreams. Freud's sources of knowledge were confined to his own experience of dreams, the dream reports of his patients, and the observation of the effects that dreams had on the psychodynamics of patients in a therapeutic setting. Allan Hobson and his commentators, by contrast, can capitalize on a huge amount of neurobiological data that have since been accumulated in the context of sleep research, much of which has been pioneered by Hobson and his colleagues at Harvard.

During the first half of the last century, the main motivation of brain research was to unravel the causes of neurological and psychiatric disorders. Hence, research focused on the investigation of the human brain and its pathologies. Due to methodological restrictions, this research was essentially confined to the postmortem analysis of structural abnormalities, and progress was slow. It was only after the Second World War that this problem-oriented approach was complemented by the strategy to search where the light is – and this light was provided by the use of animal models and the development of techniques that permitted invasive investigations of the living brain. It was now possible to establish a direct correlation between neuronal processes and associated behavior, and this paradigm shift has provided deep insights into the functional organization of nervous systems. Thanks to the more recent development of noninvasive analysis methods, this new approach can now be extended to the investigation of neuronal processes in human subjects and to relate them to mechanisms identified in animal experiments.

Following the seminal discovery of Giuseppe Moruzzi and his Pisa school in the 1950s that brain states such as sleep, arousal, and alertness are actively controlled by centers in the brain stem, sleep research became a scientific discipline. Their groundbreaking discoveries were based on the electrographic identification of brain states and their modification induced by transections performed at different levels of the brain stem and the mesencephalon. The results of these "encephale isolé" studies clearly indicated that sleep is not simply a cessation of brain functions but a sequence of controlled states characterized by distinct patterns of activity. The subsequent discovery of the so-called paradoxical sleep state, during which electrographic signals closely resemble those of wakefulness, reinforced the notion that sleep should be considered as a constitutive part of the brain's operations. However, the functions of sleep remained obscure. This is about the time when Allan Hobson and his team entered the field. In the meantime, Michel Jouvet and his school in Lyon had established close relations between paradoxical sleep episodes and the occurrence of rapid eye movements and muscle jerks. Jouvet and his colleagues coined the term "rapid eye movement sleep" (REM sleep) and thereby provided further support for the notion that this sleep phase is closely related to processes otherwise observed only in awake, highly alert brains engaged in exploratory behavior. It also became clear that not only sleep, a phase of rest, but also REM sleep is a ubiquitous phenomena found in all species endowed with complex brains.

The evidence that there is a sleep phase during which the brain exhibits all electrographic signs of alertness and generates saccadic eye movements – a motor pattern closely related to attention and exploratory behavior – suggested that this may be the phase during which dreams are generated. Experimental support for this hypothesis came from the observation that subjects consistently reported having dreamt when they were woken up during phases of REM sleep. However, dreams were also reported after phases of slow wave sleep, suggesting the possibility that dreams may also occur during these deep sleep phases. It is still an unresolved conundrum to which extent dreaming occurs during both phases of sleep and to which extent the structure of dreams differs in the two sleep stages because it is difficult to infer from reports given after awakening when exactly the remembered dream had occurred. Whether a dream is reported depends of course also on whether it is remembered. As it is likely that only a small fraction of dreams is in fact remembered – some individuals report to never dream – it cannot be ruled out that dreams occurring during REM phases are simply more easily remembered than dreams occurring during other sleep phases.

Despite the rich and bizarre phenomenology of dream contents and the prominent role of dream interpretation in psychoanalysis, sleep research put little emphasis on the investigation of putative functions of dreaming *per se*, leaving the question unanswered whether it is an epiphenomenon of a particular brain state or whether the dream, and in particular its content, once remembered, has a particular function. Sleep research rather focused on the consequences of sleep deprivation, on the neurochemical underpinnings of the various sleep stages and the gating mechanisms that increase the thresholds for sensory input and motor output during REM sleep. These studies, many of which are reviewed in detail in the three lectures by Allan Hobson and the subsequent commentaries, provide unequivocal evidence that sleep has numerous important functions and is even necessary for survival.

Prolonged deprivation of both deep sleep and REM sleep causes severe disturbances of the organism's homeostasis, impairs a host of cognitive functions and finally leads to death. Studies of the neuronal mechanisms of sleep have also revealed that the various sleep stages are controlled by a complex interplay of modulatory systems that originate in the brain stem, the pons, and the mesencephalon such as the cholinergic, the noradrenergic, the dopaminergic, the serotinergic, and the histaminergic systems. More recent studies examined the role of sleep in the context of higher cognitive functions, especially the consolidation of memories and the resolution of problems requiring insight. These investigations suggest that sleep does indeed play a role in facilitating the consolidation of procedural memory. Electrophysiological evidence indicates that characteristic neuronal activation patterns accompanying learning are replayed during early phases of sleep and that this replay may be associated with differential changes in the efficiency of synaptic connections. Whether this replay is involved in memory consolidation awaits experimental verification. Likewise, it is still unclear whether this replay is related to the observation that some of the experiences made during waking become integrated into dreams produced in the following night. By contrast, experimental support is now available for the notion nurtured by folk psychology that sleeping contributes to problem solving. When confronted with tasks requiring a considerable amount of insight in order to find shortcut solutions, the probability of finding such solutions is enhanced if subjects are exposed to the problem and then are allowed to sleep before being retested.

Self-generated, in particular rhythmic, activity plays an important role in the shaping of neuronal connections during brain development. Together with the observation that infants spend much more time sleeping than adults, this led to the proposal that the prominent oscillatory activity characterizing sleep might have an important function in supporting activity-dependent shaping of neuronal architecture during development. Another, and perhaps related, finding is that in adults the proportion of REM sleep relative to deep sleep increases following intense learning during preceding wakefulness. It has been inferred from this correlation that REM sleep might serve the rescaling of synaptic weights after they had undergone differential changes following intense learning. Synaptic connections in developing brains are particularly susceptible to use dependent modifications. Moreover, young brains are confronted with more novel stimulus material than mature brains, suggesting that developing brains are more likely than adult brains to undergo learning-dependent modifications of their architecture. Thus, the disproportionately high occurrence of REM sleep episodes in the developing brain and the increased need for sleep in general may have to do with the enhanced need for synaptic rescaling and/or memory consolidation in early life.

All these sleep-related aspects are touched upon in this book, but its main focus is on the putative functions of dreaming *per se*. In his lectures, Allan Hobson considers dreaming as an altered state of consciousness and exploits the analysis of dream states as a tool to obtain further insights into the neuronal correlates of consciousness. This is a novel and fascinating perspective as it views dreams not as an epiphenomenon of sleep-related activity but as an expression of a state of consciousness that may be considered as a precursor of the higher level consciousness manifest in the awake brain. Only a few decades ago, the proposal to link dreams with consciousness would have appeared highly speculative. The reason is that back then neuroscientists considered consciousness as a phenomenon that eluded reductionist attempts to identify its neuronal correlates. However, with the introduction of noninvasive imaging technology and the conceptual advancements in cognitive neuroscience, search for the neuronal correlates of consciousness has become a well-established field of research that already provided deep insights into the relation between neuronal processes and the various characteristics of conscious processing. In order to clearly distinguish dream consciousness from the various manifestations of awake consciousness (phenomenal awareness, self-awareness, theory of mind, rational reasoning, etc.), Allan Hobson coined the term "protoconsciousness" for the dream state. In this book, Hobson gives a comprehensive account of the features that distinguish brain states supporting dreaming from brain states supporting awake consciousness and also of the differences in the subjective experience of dreams and awake consciousness.

At the neuronal level, these differences comprise different states of the abovementioned modulatory systems and of the networks related to cognitive control and motor output. At the level of subjective experience, Hobson emphasizes differences with respect to emotional connotations, the logic of associations, cognitive control, and self-awareness. At first sight, it appears as if the meta-awareness of being aware or conscious of something appears to be reserved for states of awake consciousness, the state where cognitive control is fully expressed. However, Hobson exploits the phenomenon of lucid dreaming found in a minority of subjects to demonstrate that a trace of meta-awareness can even be retrieved in dream consciousness. Certain subjects report that they are sometimes aware of dreaming and that they even succeed to control the content of their dreams during states that are clearly identified by observers as sleeping states. Hobson proposes that these altered and intermediate states of consciousness could be fruitfully exploited to promote the understanding of conscious states in general and to instruct a specific search for neuronal correlates of consciousness. Several commentators, especially those interested in epistemic questions and philosophical approaches to consciousness, discuss the implications of Hobson's hypothesis that dream states are to be seen as protoconscious states that serve to prepare the brain for the maintenance of higher levels of consciousness both during development and maturity.

In conclusion, the data reviewed in this book and those retrievable in the scientific literature provide undisputed evidence that sleep consists of a well-organized sequence of subtly orchestrated brain states that undoubtedly play a crucial function in the maintenance of normal brain functions. These functions include both basic homeostatic processes necessary to keep the organism alive as well as the highest cognitive functions including perception, decision making, learning, and consciousness.

In view of current concepts on cognition, it is not unexpected that highly active brain states, such as occur during REM sleep, lead to self-generated patterns of neuronal activity that resemble those occurring in the awake brain during conscious processing of cognitive contents. It is commonly held that perception is based on a matching operation that compares incoming sensory signals with hypotheses, so-called priors, that correspond to activated memory traces stored in the brain's functional architecture. This notion agrees well with data from non-invasive imaging which indicate that imagery of a cognitive object is associated with spatiotemporal patterns of activity that closely resemble those generated when the respective object is actually perceived. Likewise, such investigations revealed that hallucinations go along with activation of exactly those brain structures that would also become active were they engaged in processing real stimuli corresponding to the hallucinated content. Thus, evidence indicates that strong, self-induced activation of cortical networks can lead to vivid experiences because the resulting patterns are similar to those occurring during perception of external stimuli. Such states are observed in a variety of pathological conditions such as migraine, fever, epileptic seizures, and psychosis, but at least in terms of electrographic signatures seem to occur also in REM sleep. In the latter, these self-generated activation patterns are not constrained by sensory signals from the outside world because during this sleep phase the sensory gates are closed.

One might speculate that it is this lack of constrainment that allows for the bizarre associations characteristic for dreamt contents. The evidence that only a fraction of dreams seems to be remembered might suggest that nature has implemented mechanisms that prevent memorization of dreams or at least prevent access of the dreamt material from conscious recall in the wake state. The fact that the aminergic systems are down regulated during REM sleep may be one reason for dream amnesia, as these systems are involved in gating synaptic mechanisms involved in memory formation. Thus, one might consider the possibility that nature has not only implemented mechanisms that decouple the sleeping brain from sensory input and motor output but also prevent the sleeping brain from becoming aware of the self-generated activity patterns. If so, one might speculate that these three processes serve to protect the sleeping brain from waking up prematurely. If this were the case, one would have to assume that awakening in the middle of the night as a result of strong dream emotions would simply indicate that the suppressive mechanisms have been overridden. Likewise, the increase of remembered dreams towards the end of a normal night would then be the consequence of the gradual increase of activity in the aminergic modulatory systems that precedes awakening and prepares the transition from sleep to wakefulness and the gradual recovery of cognitive control. This scenario would also be compatible with the indications that lucid dreaming occurs preferentially in this transition period.

If one adhered to this interpretation, one would have to conclude that highly evolved brains require periods during which they can activate internal representations unconstrained by sensory input and that nature has implemented mechanisms that protect these states from being interrupted by premature awakening that would result if the self-generated activity would trigger motor responses or have access to the level of conscious awareness. According to this interpretation, awaking from nightmares and remembering dreams would simply be a reflection of incomplete elimination of dream contents from conscious awareness. This in turn would imply that becoming aware of dream contents and remembering them has no adaptive function. If this were the case, it would, however, by no means imply that dream contents, if they penetrate into awareness and are remembered, have no impact on brain functions and are devoid of interpretable content. Those who remember dreams know only too well how strongly they can influence the cognitive and emotional dispositions in the following wake period, especially when they have been loaded with strong emotions. Likewise, the contents recombined in remembered dreams often do not only reflect fragments of experiences made during the preceding day but often recombine motifs that have distinct emotional connotations of biographical relevance. This cannot be otherwise since it is the dreamer's brain that generates these unconstrained and sometimes remembered activation patterns and it is the dreamer's brain that determines which of these patterns will penetrate into the workspace of consciousness and be remembered.

Prof. Wolf Singer
Max Planck Institute for Brain Research,
Frankfurt Institute for Advanced Studies (FIAS),
Ernst Strüngmann Institute (ESI) for Neuroscience
in Cooperation with Max Planck Society,
Frankfurt am Main, Germany

Acknowledgments

I am very grateful to Paul O'Pree and Jane Broadbent for their invitation, sponsorship, and administration of the William James Lectures given at the University of Roehampton in March 2008. It was there that I became aware of the residual Freudianism against which my theory is cast. To Professor Friedrich Stadler and his staff in Vienna, I am grateful for the opportunity to present and to publish this work. The collection of the commentaries and the production of the manuscript are the work of Nicholas Tranquillo, my longtime assistant and friend. The discussants responded to the ideas of these lectures with great enthusiasm, for which I am deeply appreciative. I am grateful to David Borodin and Jay Vogelsong for permission to publish their works of art.

J. Allan Hobson

Contents

Biographies

J. Allan Hobson (author) is an Emeritus Professor of Psychiatry at Harvard Medical School, where he conducted research on sleep and dreaming as the Director of the Laboratory of Neurophysiology at the Massachusetts Mental Health Center. This book represents his thinking about human consciousness as expressed in the William James Lectures at Roehampton University. He has written 12 previous books and published over 200 original scientific articles. He indulges his interests in art and architecture at his US home in East Burke, Vermont, and with his Italian family in Messina, Sicily.

Nicholas Tranquillo (editor) has been associated with Allan Hobson since 1997, when he was a student in Psychology E1450, The Biopsychology of Waking, Sleeping and Dreaming. Over the intervening years, he has served as a manager of the Laboratory of Neurophysiology at Harvard Medical School. In addition, he has been deeply involved and responsible for manuscript production, having worked on six successful books. In the preparation of the present volume, he solicited and edited the commentaries as well as taking full responsibility for readying the book for publication. Complementing his academic interests are commitments to culinary creation, remote and exotic travel, Qing Dynasty research, and the love of opera. He lives in Salem, New Hampshire.

Contributors

John Antrobus City College of New York, New York, NY, USA

Mercedes Atienza Laboratory of Functional Neuroscience, Spanish Network of Excellence for Research on Neurodegenerative Diseases (CIBERNED), University Pablo de Olavide, Seville, Spain

Susan Blackmore School of Psychology, University of Plymouth, Plymouth, UK

Jan Born Department of Medical Psychology and Behavioral Neurobiology, University of Tübingen, Tübingen, Germany

Allen R. Braun National Institute on Deafness and Other Communication Disorders, National Institutes of Health, Bethesda, MD, USA

José L. Cantero Laboratory of Functional Neuroscience, Spanish Network of Excellence for Research on Neurodegenerative Diseases (CIBERNED), University Pablo de Olavide, Seville, Spain

Vanessa Charland-Verville Coma Science Group, Cyclotron Research Centre and Neurology Department, University of Liège, Liège, Belgium

Michael Czisch Max Planck Institute of Psychiatry, Max Planck Society, Munich, Germany

Armando D'Agostino Section of Psychiatry, Department of Medicine, Surgery and Dentistry, University of Milan Medical School, Milan, Italy

Antonio Damasio Brain and Creativity Institute, University of Southern California, Los Angeles, CA, USA

Daniel C. Dennett Center for Cognitive Studies, Tufts University, Medford, MA, USA

Susanne Diekelmann Department of Neuroendocrinology, University of Lübeck, Lübeck, Germany

Martin Dresler Max Planck Institute of Psychiatry, Max Planck Society, Munich, Germany

Michael W. Eysenck Department of Psychology, Royal Holloway College, University of London, London, UK

Department of Psychology, University of Roehampton, London, UK

Dominic H. Ffytche Department of Old Age Psychiatry, Institute of Psychiatry, King's College London, London, UK

Karl Friston The Wellcome Trust Centre for Neuroimaging, Institute of Neurology, London, UK

J. Allan Hobson Department of Psychiatry, Harvard Medical School, Boston, MA, USA

Ludwig Huber Messerli Research Institute, University of Veterinary Medicine, Vienna, Austria

John F. Kihlstrom Department of Psychology, University of California, Berkeley, Berkeley, CA, USA

Don Kuiken Department of Psychology, University of Alberta, Edmonton, AB, Canada

Steven Laureys Coma Science Group, Cyclotron Research Centre and Neurology Department, University of Liège, Liège, Belgium

Joseph LeDoux Center for Neural Science, New York University, New York, NY, USA

Rodolfo R. Llinás Department of Physiology and Neuroscience, NYU Medical School, New York, NY, USA

Thomas Metzinger Philosophisches Seminar, Johannes Gutenberg-Universität Mainz, Mainz, Germany

Andrea Moro Center for Neurolinguistics and Theoretical Syntax (NeTS), Institute of Advanced Study, IUSS, Pavia, Italy

Adrian R. Morrison Department of Animal Biology, Laboratory for the Study of the Brain in Sleep, School of Veterinary Medicine, University of Pennsylvania, Philadelphia, PA, USA

Tore Nielsen Département de Psychiatrie, Université de Montréal, Montréal, QC, Canada

Valdas Noreika Department of Psychology, Centre for Cognitive Neuroscience, University of Turku, Turku, Finland

Department of Psychiatry, Oxford Centre for Human Brain Activity (OHBA), Warneford Hospital, University of Oxford, Oxford, UK

Edward F. Pace-Schott Department of Psychology, University of Massachusetts, Amherst, MA, USA

Michael Posner Institute of Cognitive and Decision Sciences, University of Oregon, Eugene, OR, USA

Allan Rechtschaffen Departments of Psychiatry and Psychology, College University of Chicago, Chicago, IL, USA

Antti Revonsuo School of Humanities and Informatics, University of Skövde, Skövde, Sweden

Department of Psychology, University of Turku, Turku, Finland

Silvio Scarone Section of Psychiatry, Department of Medicine, Surgery and Dentistry, University of Milan Medical School, Milan, Italy

Michael Schredl Central Institute of Mental Health, Mannheim, Germany

Wolf Singer Max Planck Institute for Brain Research, Frankfurt Institute for Advanced Studies (FIAS), Ernst Strüngmann Institute (ESI) for Neuroscience in Cooperation with Max Planck Society, Frankfurt am Main, Germany

Mark Solms University of Cape Town, Cape Town, South Africa

Arnold Pfeffer Center for Neuro-Psychoanalysis, New York Psychoanalytic Institute, New York, NY, USA

David Spiegel Stanford University School of Medicine, Stanford, CA, USA

Victor Spoormaker Max Planck Institute of Psychiatry, Max Planck Society, Munich, Germany

Giulio Tononi Department of Psychiatry, University of Wisconsin, Madison, Madison, WI, USA

Claus Vögele Unité de recherche INSIDE, Université du Luxembourg, Walferdange, Luxembourg

Ursula Voss Universität Bonn, Bonn, Germany

JW Goethe-Universität, Frankfurt, Germany

Renate Wehrle Max Planck Institute of Psychiatry, Max Planck Society, Munich, Germany

Lawrence Weiskrantz Department of Experimental Psychology, University of Oxford, Oxford, UK

Jennifer M. Windt Philosophisches Seminar, Johannes Gutenberg-Universität Mainz, Mainz, Germany

Hellmut Wohl Boston University, Boston, MA, USA

Part I
The William James Lectures on Dream Consciousness

Chapter 1
Introduction

J. Allan Hobson

In March of 2008, I gave the following three lectures at the University of Roehampton in London, England. The invitation, from the Vice-Chancellor, Paul O'Pree, came in response to my presentation, the previous year, of a summary of the ideas in the book, *Angels to Neurones* (subtitled *Art and the New Science of Dreaming* (Hobson and Wohl 2005)), written in collaboration with Hellmut Wohl. The text of the lectures was submitted, in advance, to Roehampton and circulated to its faculty for commentary. Of the nine responses received, five were very negative. These negative critiques came from a wide variety of liberal arts scholars all of whom were committed, in one way or another, to Freudian psychoanalysis. To me, this fact signaled the degree to which Freudian psychological theory, which I consider to be erroneous and obsolete, was embedded in the modern academic mind.

Many of the ideas in these lectures ripened and fueled my Nature Reviews Neuroscience article (Hobson 2009), published in November, 2009, entitled "REM Sleep and Dreaming: Toward a Theory of Protoconsciousness." My very helpful editor at Nature, Leonie Welberg, attended the Roehampton lectures and was, like me, astonished at the vituperative response of my critics. The scientific reception at Roehampton and around the world to the protoconsciousness hypothesis was, and remains, contrastingly appreciative. It would seem that C.P. Snow's *Two Cultures* paradigm is still valid (Snow 1960). Scientists and humanitarians have difficulty talking to each other. This is regrettable, especially since the protoconsciousness hypothesis has many important implications for the humanities.

Hoping to make further inroads into the art-science hinterland, I responded with enthusiasm to Friedrich Stadler's invitation to give a lecture in the context of his Vienna Circle Summer School Program in July of 2010. I have always felt that William James' triumvirate of Psychology, Physiology, and Philosophy held a key to unlocking interdisciplinary dialogue. A bridge between brain science and the humanities might well be built upon the foundation of two legs of the Jamesian triad

J.A. Hobson (✉)
Department of Psychiatry, Harvard Medical School, Boston, MA, USA
e-mail: allan_hobson@hms.harvard.edu

N. Tranquillo (ed.), *Dream Consciousness*, Vienna Circle Institute Library 3,
DOI 10.1007/978-3-319-07296-8_1, © Springer International Publishing Switzerland 2014

even if neurobiology could not yet reach certain proponents of psychology. The enthusiastic reception by philosophy faculty and students to my lectures in Vienna told me that this intuition might be correct and the publication of my Roehampton William James lectures, illustrated by some of my Nature Neuroscience figures, became an idea that was attractive to me. If the lectures and commentaries of understanding colleagues could accompany my writings, the integration that I was hoping for at Roehampton might finally be achieved or at least enhanced.

1.1 The Protoconsciousness Hypothesis

For most of the twentieth century, dreaming was seen as an unconscious mental process. In these lectures, I will attempt to show that it is both possible and profitable to regard dreaming not as unconscious but rather as an altered state of consciousness that is difficult to recall in waking. Dreaming is a state of consciousness that proceeds in sleep without awareness of the outside world; its content is difficult to remember but rich; and when it is remembered it seems strangely different from waking consciousness. Regarded as unconscious mental activity, dreaming has thus invited interpretation for centuries and dream interpretation has impeded science until quite recently. Particularly egregious is Sigmund Freud's *Interpretation of Dreams* (1900), which still exerts an antiscientific force on our understanding of dreaming.

Regarding dreaming as an altered state of consciousness rather than an unconscious mental process is legitimate because each of the modules of waking consciousness changes in some specifiable way in dreaming. We can thus assert, with confidence, that consciousness is present as dreaming during sleep but that it is in an altered state. In these lectures, I use the word "altered" in a strictly naturalistic sense without meaning to invoke any new-age implication and without reference to drug taking. According to me, dreaming is thus not so much unconscious as it is unremembered. The physiological reasons for the occurrence of dream consciousness–and for its isolation from waking consciousness–are now well understood, as will be summarized in Lecture II.

This change in emphasis from dreaming as an unconscious state to dreaming as a conscious state is profitable because it permits an analysis of dreaming in terms of increasingly well understood brain mechanisms that mediate the psychological modules that make up the experience of dreaming. From such an analysis, it is immediately apparent that dream consciousness can be compared to waking consciousness as a way of understanding not only dreaming but the evolution of consciousness itself. It is hoped that this new approach may make a contribution to our understanding of the mind in relation to mental illness and to the solution of the otherwise intractable mind-body problem.

My strategy in approaching this problem scientifically is laid out in Table 1.1. As can be appreciated by reference to the table, dream science integrates data from a variety of techniques in humans and animals. The collection of (objective)

Table 1.1 Methods of dream neuroscience

Human subjects	Animal models
Collect and analyze dream reports from journals and tape recordings	Not available
Record polygraphic sleep in laboratories and at home	Record polygraphic sleep Experimental lesions Electrical stimulation Microelectrode for cells Microinjection of chemicals
Image brain during sleep using MRI, PET, SPECT	Image brain during sleep using MRI, PET, SPECT

third-person data is a check and a constraint against the unreliability of first-person (subjective) reports. The universality of formal dream features in humans and the homologous nature of sleep in most mammals give the enterprise scientific solidity.

As will be stressed in Lecture 1, scientific dream research begins with the collection and analysis of subjective reports. The data from these reports are correlated with physiological recording data in a sleep lab or at home using portable devices. Hence, there can be mapping from human mind to human brain and from human brain to animal brain. This approach assumes homology between humans and other mammals at the level of the subcortical brain (the source of REM and primary consciousness).

In animal models, more-detailed information about the neurophysiology is obtained through lesions and brain stimulation, microelectrode single cell recording and neuropharmacological microinjection techniques. These data can be correlated with polygraphic sleep data and inferences made about human physiology that might explain the observed differences in psychology that characterize waking, sleeping, and dreaming.

Brain imaging can be used in both human and animal subjects to detect regional activation patterns associated with sleep physiology and dream psychology, although to date, there has been little brain imaging study of animal sleep. This is unfortunate because it is from animal models that we obtain the cellular and molecular level evidence on which theory building depends.

An important upshot of the new theory of protoconsciousness is that REM sleep dreaming is an automatic brain-mind process that provides a structural and functional building block for waking consciousness. The two states interact dynamically over the life span in such a way as to enrich both of them. Dream consciousness guarantees the binding of sense of self, motility, sensation and emotion. It is upon this base that waking consciousness is built.

The fact that a REM sleep-like state precedes waking in fetal and neonatal development is an important argument in favor of the theory. A cause must precede an effect but of course precedence does not prove causation. Another important biological fact in favor of the theory is that REM sleep has developed twice in evolution: both mammals and birds evince REM. It is notable that both mammals and birds have relatively large and/or complex brains that mediate their impressive behavioral and cognitive repertoires. Moreover, both mammals and birds are uniquely homeothermic, possibly in keeping with the need to regulate the temperature of their complex brains. Finally, and most strikingly, the capacity for homeothermy is REM sleep dependent. Like antecedence in time, the plausibility of correlated biological attributes is a strong but inconclusive argument for the validity of the theory.

It makes good sense to prepare the brain for subsequent waking and the physiology of REM sleep manifests abundant signs of an activation state favoring the massive and parallel connectivity essential to the binding of the multiple cognitive processes that are required to explain the extraordinary unity of waking consciousness. For this reason, I refer to REM sleep dreaming as a proto-conscious state. By protoconscious, I mean both prior to and fundamental to waking consciousness.

1.2 Testing the Hypothesis

Definitive proof of the protoconsciousness hypothesis may be as difficult to obtain as definitive evidence for Charles Darwin's theory of evolution (Darwin 1988). This qualifier is not as much meant to flatter protoconsciousness theory as it is to admit that crucial experimental tests of the theory may be difficult to conceive. One reason for this is that the deprivation paradigm is morally unacceptable when applied to human beings. The theory is nonetheless useful as a catalyst for the more searching and critical scientific exploration of the development, realization, and maintenance of the human brain mind: three areas of science which are currently under-investigated. For a small university like Roehampton and a special purpose institution like the Vienna Circle, the opportunity to contribute to the science of dreams is as real as the chance to examine the impact of the new science of dreams on the great variety of humanistic disciplines already well established in academia.

The advocates of pre-ordained intelligent design may prefer to interpret dreams as prophecy in the tradition of the Holy Bible and Freud's secular bible, *The Interpretation of Dreams*. But it is already clear that such an approach is both inadequate and largely erroneous. This opens the door wide to an alternative formulation.

In these three lectures, I will expose the new psychology (Lecture I), the new physiology (Lecture II), and the new philosophy (Lecture III), that derive from and feed the protoconsciousness hypothesis. The lectures are named for William James in recognition of his insistence that psychology, physiology, and philosophy be unified (James 1902; James 1981). In so far as his philosophy touched upon even such remote fields as theology, the Jamesian tradition of tolerance and inclusiveness is also honored, although I do not pretend to be as broad-minded as James.

My main goal here is to make clear a new theory rather than cast doubt on old ones. Of course, science itself has a winner-take-all structure but, by wanting to become one of the humanities, the firearm of brain science must be left outside the tavern of academic integration. I will therefore try to be as gentle as possible as I develop my story. This effort involves suppression of my own personal disappointment with Freudian psychoanalysis, a theme which is treated in more detail in *Dream Life: An Experimental Memoir*, published by MIT Press in 2011 (Hobson 2011).

1.3 The Commentaries

When Friedrich Stadler indicated that he might be willing to publish my lectures, it occurred to me that this was a marvelous opportunity to seek commentary from a wide range of colleagues. I therefore invited them by asking each to answer a specific question for me. This is a variation on the theme of open peer review that has made the Behavioral and Brain Sciences journal so justifiably popular. The response was gratifyingly broad and deep. It is a pleasure to share them with you as well as to respond to their imaginative and critical content.

In fairness to Sigmund Freud, he deserves a debt of gratitude for expressing his own theory of dreams sufficiently clearly that it could be refuted by me and others.

References

Darwin, C. (1988). The Origin of Species, 1876. New York: New York University Press.

Freud, S. (1900). The interpretation of dreams (Translated from German and edited by J. Strachey). New York: Basic Books.

Hobson, J. A. (2009). REM sleep and dreaming: Towards a theory of protoconsciousness. *Nature Reviews Neuroscience, 10*, 803–813.

Hobson, J. A. (2011). *Dream life: An experimental memoir*. Cambridge, MA: MIT Press.

Hobson, J. A., & Wohl, H. (2005). *Angels to neurones: Art and the new science of dreaming*. Parma: Mattioli. 1885.

James, W. (1902). *The varieties of religious experience; a study in human nature; being the Gifford lectures on natural religion delivered at Edinburgh in 1901–1902, by William James*. New York: Longmans, Green.

James, W. (1981). *The principles of psychology*. Cambridge, MA: Harvard University Press.

Snow, C. P. (1960). *The two cultures*. Cambridge: Cambridge University Press.

Chapter 2
Lecture I: Psychology

J. Allan Hobson

Proof That Dreams Are Real

First, take a brain and steep it in a world.
Then, lay it by a while to set. Next, go
Reduce the voltage till the waves unfurled
Are quicker paced and nothing's felt below.
Now, activate some cells, unlocking stored
Emotions and events as they'd appeared,
And watch the nonsense loosed excite a score
Of pathways making these stray bits cohere.
Yet, this state dreamt was but a new event;
It "happened" like those "real" ones we'd espied—
Remembered, from the world—yes, underwent
As if it too were input from outside.
Thus, the perceived and dreamed, seen close enough,
Reveal that they're both made of that same stuff!

–David Borodin

Despite the difficulties with recall, most of us can remember at least a few dreams vividly. Our dreams always have certain features in common, as I will try to illustrate with an example from my own dream consciousness experience. I hope this exercise will be both enlightening and inspiring. The Holy Grail of dream science may be in your own bedroom.

J.A. Hobson (✉)
Department of Psychiatry, Harvard Medical School, Boston, MA, USA
e-mail: allan_hobson@hms.harvard.edu

N. Tranquillo (ed.), *Dream Consciousness*, Vienna Circle Institute Library 3,
DOI 10.1007/978-3-319-07296-8_2, © Springer International Publishing Switzerland 2014

2.1 London Tree Trimming Dream-Verbatim Report

I am in London. I am riding a bicycle in traffic, toward an uncertain location at which I am going to trim a tree. To that end I am carrying across the handlebar of my bicycle a tree trimmer tool, about six feet long, that has, at one end, a hardware device for culling an out-of-reach branch and, by pulling a rope connected to a moveable blade, cutting it off without a ladder or pruning shears.

I assume that I am in the right part of the town when I notice that the architecture is typical town house style of Georgian England (as seen on Wimpole Street or on Beacon Hill in Boston). It is in such urban neighborhoods that people may still have trees, small gardens, and shrubs like the one I suppose I will trim.

The house I stop at has a very small tree in front of it but is hardly worth trimming. There is certainly no need for the long-handled apparatus that I have brought to the job. A young girl in 18th century dress comes out of the house, making me feel that I am at the right place after all.

I therefore ask her to call her father so that we can discuss the goals and terms of the tree trimming job. An elegantly frock-coated man appeared and smiled at me as if he were my employer.

There are two ways to approach this account. The interpretative approach seeks to understand the dream in terms of its autobiographical psychodynamic significance. What does the tree-trimmer symbolize? Why do I take on such foolish assignments? Why am I always so over prepared for my work routines? Why do I persist when all the evidence is against my success? (Fig. 2.1)

These are perfectly reasonable and relevant questions which would lead to a reasonable and relevant hypothesis about why I am the way I am. I could, for example, record my associations to the dream material such as the tree pruner and show that they lead to explicit and tense interactions with my father who was obsessed with gardening and landscape.

2.2 The Formal Analysis of Dreams

Whatever one's reaction to the interpretative approach, it can safely be asserted that no attention is paid by most dream interpreters to the alternative and complimentary approach of formal analysis. In order to understand the difference between dream content analysis (by what I call the interpretative approach) and dream form analysis (by what I call the mental status approach) we need only consider the difference in clinical psychology and psychiatry between taking an autobiographical history and performing a formal mental status exam.

In a case like that of my 80 year old mother, the history concerned her getting lost in familiar environments; her not knowing the names of people familiar to her; and her intermittent confusional states such as not knowing what she was doing or why she was doing it. The mental status exam revealed disorientation (especially for time and place but also for persons), memory loss (especially for recent events), and confabulation (making up stories to cover the holes in her memory).

<u>DREAM REPORT</u> I drew and recorded these dream recollections in Vol. 51 of my personal journals on 10/2/1990. Each of the five subsequent squares contains a drawing and the relevant part of the report.	(1) I am riding my bicycle in London. Across the handlebars is a 6 ft. tree trimmer. I assume I have an appointment to trim a tree. 
(2) But I am not sure of my destination. Then I realize that my destination is 18th century London. e.g. Wimpole Street, Row houses etc. 	(3) I arrive at a house in 18th century London. There is a girl wearing an 18th century dress. I assume that the neighborhood is having a period dress festival. 
(4) I see a small tree in the yard and assume I am to trim it but it is only 4 feet high and very scrawny. 	(5) A man with mutton chop sideburns and 18th century frock coat appears. He is jovial and smiles broadly at me. I assume he is the one who hired me. ...I wake up 

Fig. 2.1 London tree trimming dream report

The fact that my mother had Alzheimer's disease had little to do with her biographical history. It was, rather, biology that determined the symptoms. Her brain cells had degenerated to the extent that she had state-dependent psychological problems. The essence of the mental status concept is that whatever its content, the mental state of a person reflects that person's brain-state. By means of the formal analysis of dreams, we may thus be able to use the mind to study the brain more or less directly. I am already jumping to a startling conclusion: what we call mind is in fact a brain function. There is no deeply dualistic assumption lurking here. As discussed in greater detail in Lecture III, dual-aspect monism holds that REM sleep and dreaming are two sides of the same coin. One side is subjective; the other side is objective. Thus, I use the hybrid term *REM sleep dreaming* to refer to a unified state of the brain-mind.

My tree trimming dream has formal features which are every bit as telling as my mother's response to certain aspects of her mental status exam. Her symptoms were perhaps more easily understood because they were so flagrant. But let us reflect for a moment. Her symptoms occurred in the context of waking consciousness. Suppose that my dream perceptions, feelings, and thoughts had occurred in waking. What I would have made of them then is quite a different story, you must agree. Let us walk through the mental status exam and see what it shows about my dream.

1. *General Appearance and Behavior*

   In the dream, I am me. I don't see myself as a third person, hence I do not know what I looked like nor do I hardly ever see myself in my dreams whereas the monitoring of my appearance is an important aspect of my waking consciousness. Self-reflective awareness is therefore as markedly deficient in my tree trimming dream, as it is in almost all of my dreams.

   In my recent dreams, the fact that I am often riding a bicycle (or skiing or swimming) is surprising since I have not been able to ride a bicycle in the 7 years since a brain stem stroke made any balanced motor activity impossible. So, already, in my personal attributes, there is a marked discrepancy between the waking me and the dreaming me. At no point in any of my dreams does it seem peculiar that I am riding a bicycle. In my tree trimming dream, that impossible behavior goes off without a hitch.
2. *Stream of Talk*

   Compared to a face-to-face interview in waking, very little is said when I am dreaming. I do talk to myself a bit as I reassure myself about my dream goal and destination. And when I get there I do ask the girl to go and get her father. I have had dreams with more discourse than this but I often find that my communications (and even my hearing of music) are soundless, as if the perception and emotion were stimulus independent, which of course they are. Dreamers often report that they "just knew" what was said or heard.

   Very recently I have begun to talk in my sleep. I often give lecture-like speeches heard only by my wife and others within earshot of my bedroom. Just the other night I addressed a group of students about writing instruments. According to my stepson, Luca, I said: "I never use pencil. Fountain pens are too messy and ballpoint pens are too crude. The best pen is architect's fine point soft tip." Somewhere around "ball point pens" my own voice woke me up. I was surprised to hear myself talking so clearly while asleep and listened, with interest, to the rest of the speech, which felt just as automatic as are my over-learned lectures. As far as I know, this was the first and last time that I addressed the subject of writing instruments in a dream or any other state of consciousness. But the opinions that I expressed are sincere and deeply held by me.

Since writing this chapter, I have been diagnosed with classic sleep apnea. My treatment consists of using a continuous positive airway pressure machine (CPAP). With CPAP operating, I have stopped lecturing in my sleep!

I wasn't dreaming when I gave that sleep talk and I am not saying that dreaming can't be loquacious. I assert only that dream talk is rare but that, when it is heard, it can be quite clear. Speech delivery, when it is permitted, is just as likely to be lucid (as in my felt-tipped pen discourse) as it is to be garbled (as is more typical of the sleep talking that occurs earlier in the night). Verbal receptivity is, if anything, enhanced in dreams so that communication sometimes takes place without any speech whatsoever. This process is similar to thought broadcasting and the delusional idea that people are talking about a schizophrenic patient. Dream speech is in need of much more careful study.

3. *Clarity* versus *Clouding of Consciousness*

I now feel that my dreaming consciousness is just as clouded as was my mother's in her confused states. The sharpness of perception and the sense of realistic clarity are at the same time enhanced. Dreaming is truly surreal as Leonardo da Vinci's rhetorical question suggests: "Why does the eye see a thing more clearly in dreams than it does when awake?" One answer, of course, is that the eye doesn't see at all in either state. The brain sees whatever image the eye, or some other part of the brain, sends it.

In waking, the eye and its brain (that is to say the retina) are essential to the perception of external stimuli. But when I am asleep, my eyes are closed and it is dark. When I dream so convincingly of riding my bike through London, you could say that my vision of London is less detailed than it would be if I was really seeing those distinctive red phone booths, the unique double decker buses and the quaint black hackney cabs, but my consciousness of London is nonetheless just as keen in my dream as is Shakespeare's awareness of his lover in his dream sonnet 43. So keen is my sense that I am in London that I do not notice the absence of local place markers. I don't need them because my perceptual consciousness in enhanced. This belief in false private perception is exactly what happens when people go crazy (see the subsequent discussion of hallucinations and delusions).

4. *Intellectual Functions*

Knowing how to read and calculate are important cognitive skills which I, for one, practice a lot in my waking consciousness. But I hardly ever do so in my dreams. My London tree trimming dream is typically devoid of either reading or calculations. I do not really know where I am going, but I do not seek any street signs. And I do not succeed (or fail) to read them as I would were I awake. And I do not wonder how much I will charge or be paid for my tree trimming services as I would most certainly do if I was awake. When I see the too small tree, I don't say to myself: "I will charge this guy an hourly rate of X just for showing up." There is no X and I don't notice this very unprofessional lapse.

If I had been awakened from this dream I would have been able to read and find my way around on a map of London. I could easily subtract 7 from 100 and get 93 and subtract 7 again and get 86, then 79 and so on (as long as I was not sleep deprived, which would make even simple math difficult). I would know that George Bush was at that time still the president of the U.S. and that Barack Obama and Hillary Clinton were then battling for the democratic presidential nomination. But current events play no more part in most of my dreams than do math or reading. I may often puzzle over some problem in my own private life, like why I am in London when I suppose I am in Boston. In brief, when I dream my mind doesn't function very well as an analytic engine.

5. *Orientation to Time, Place and Person*

I know who I am even if the me I take myself to be has (or lacks) attributes that are not current (like dream consciousness bike riding when I am incapable of wake consciousness bike riding), I could be in London (because I am anticipating going there) but I am in fact asleep in my bed in Boston. My confidence about my location is not as bad in this particular dream as it sometimes is. Many times when dreaming, I may feel as confused as my poor Alzheimer's diseased mother when she had no idea where she was. She may still think she is the very sane Anne Hobson, and be able to tell you so even though she has no idea who you are or where it is that you are conversing with her. By clever feigning and confabulation she may fool you into thinking she is really ok. Just as you and I fool our critical selves into thinking everything is ok in dreams when we are riding a bike with a tree trimmer across its handlebars through what we take to be downtown London. The date, as usual, is unspecified in my London tree trimming dream and I do not look at my dream watch to see how I am doing in biking to a job interview. Time is rarely specified by dream consciousness. We seem only to know that the time is now, whatever time that may be. This formal dream feature has not yet been scientifically studied.

In other words mental functions so essential and prominent in anchoring waking consciousness to the here and now are markedly deficient in dreaming. Strangely enough this gross cognitive defect, so strongly suggestive of organic brain dysfunction, had escaped the notice of dream interpreters until very recently. This is surprising since artists, like the surrealist painter, Salvador Dalí, have long recognized formal dream features and represented them in their work. Table 2.1 lists the formal dream features which have, so far, been scientifically studied.

I am reminded of my patient, Irvin Monroe, whose flagrant paranoia seduced my psychodynamically-tuned mind so strongly that I forgot to ask him if he knew what day it was. As his long-suffering sister pointed out to me, he *didn't* know what day it was. He was also disoriented because his brain was soaked in alcohol, but it was the sub-arachnoid hemorrhage that killed him. I missed that diagnosis because I was too interested in Irvin's history and in psychoanalytic theory to perform a simple, routine, mental status exam. Perhaps it is embarrassing mistakes like that which impel me not to miss the neurological

Table 2.1 Some formal properties of dream consciousness of relevance to neurobiology. These studies, which are summarized, were all performed in the Laboratory of Neurophysiology, Harvard Medical School. Boston, MA

Phase I (1980–1994): Laboratory and unmonitored home reports	
Perception	Vision and sense of movement predominate; pain and taste are rare
Bizarre cognition	Times, places, and persons change without notice; (bizarreness is measured as plot discontinuity and incongruity)
Fantasy	Chimeric characters are common in dreaming but absent in fantasy
Children	Adult type dreaming begins ~ age 5
Emotion	Anxiety (fear), elation, and anger predominate; sadness, guilt, and depressed affect are rare
Plot sequence	Gradual loss of orientation within scenes; radical loss of orientation across scenes
Splicing	Judges cannot recognize continuity across scenes (Hobson et al. 1987; McCarley 1982; Merritt et al. 1994; Resnick et al. 1994; Stickgold et al. 1994a, b; Sutton et al. 1994a, b; Williams et al. 1992)
Phase II (1995–2004): Home dreams with physiological monitoring	
Report length	REM reports 7× longer than NREM reports
Sensation of movement	More common in REM than NREM reports
Character recognition	Unreliable in REM but dreamer does not notice errors
Thinking	Highest in waking, lowest in REM; reciprocal with hallucinating across states
Memory source	Identified in only 20 % of dream incidents (80 % of dream events synthesized de novo) (Fosse et al. 2001, 2002, 2003; Hobson et al. 2000; Kahn et al. 2000, 2002; Pace-Schott and Hobson 2002; Porte et al. 1996; Stickgold et al. 2000; Walker et al. 2002, 2003)
Phase III (2005–present): Home dreams with physiological monitoring – focus on secondary consciousness (metacognition)	
Theory of mind	Dreamer recognizes mental process of other dream characters
Logic	Some dream thinking is rational; most dream thinking is non – rational
Authorship	Dream reports incorrectly grouped by judges
Schizophrenia	Patients and controls have equally bizarre dreams; patients, but not controls, have bizarre TAT responses (TAT = Thematic Apperception Test) (Hobson et al. 2011; Kahn and Hobson 2005)

diagnosis of dreaming. We probably could not have done much for Irvin Monroe but if we do a mental status exam of dreaming we can very possibly save a whole field of human inquiry from missing the brain basis of consciousness.

6. *Mental Content*

This part of the mental status exam interests us psychiatrists most because it asks the subject about internal perceptions (hallucinations) and false beliefs (delusions). What this part of the mental status exam does is to assess psychosis, the degree to which the subject is out of contact with reality. When I dream of riding my bicycle in London traffic on my way to a tree trimming job, I am disoriented to a psychotic degree because I can actually see and feel myself to be in London when I am in fact in bed in Boston. My movement through the dream space is smooth, my vision of the traffic and buildings is continuous and convincing (or even surreal as I have already pointed out). There is thus no room for doubt. The hallucinated perception of bicycling through London dominates my mind. There is no room for doubt, and anyway, no one at home to do the doubting.

Allan Rechtschaffen has astutely characterized this aspect of dream consciousness as "single-mindedness" (Rechtschaffen 1978). Not only am I in the grips of both hallucination and delusion but I have no alter ego to help me out of this mental mess. Self-reflective awareness, background-foreground monitoring, insight and judgment, all dissolve with this loss of my second self, my observing ego. This cognitive failure has led me to suspect that my dreams have me as much as I have the dreams!

Protoconsciousness theory holds that the dreaming brain, activated in sleep, produces a self, an I, who is doing something (riding a bicycle), somewhere (London), toward some end (trimming a tree). All this is fabricated but it is a scenario. My brain, when activated as it is in REM, is therefore a scenario generator. The same system might well support the scenario structure of my waking consciousness when the external reality of the world supplies the details of time, place, other persons, and a goal for my behavior. In other words, I am suggesting that the brain, whenever and however it is activated, generates a scenario structure for consciousness. This is quite useful, especially if we can reliably suppress the tendency to dream when we are awake. Unfortunately, not everyone can do this. Some of us become "mentally ill" instead!

Silvio Scarone and his fellow psychiatrists at the Ospedale San Paolo in Milano have shown that psychotic patients have the same amount of bizarreness in their dreams as do normal people (Scarone et al. 2008). Both patients and normals are thus equally "crazy" when they dream. But while normal people control and contain bizarreness during their waking consciousness, psychotic patients do not. Their waking consciousness is just as bizarre as their dreams. They also lack the self-reflective awareness, the insight and judgment, and what we call perspective in social awareness. This is a severe handicap, as are the other cognitive defects summarized in Table 2.1.

7. *Dream Memory*

Waking consciousness has access to a vast domain of accurate details of past knowledge and past experience. Memories can be explicit and semantic (when they convey facts and knowledge) or implicit and procedural (when they underlie skills and know-how). It can be argued that memory is our most important mental power. By means of calling up facts and skills from memory, our cognitive and behavioral repertoires are enhanced immeasurably. Of course, most of the information that is stored in our memory is non-conscious until it is called or kicked up into consciousness by an association or by our command.

We should not assume that our occasional failure to access information stored in the brain is due to active interference with retrieval. We don't yet know enough to be sure that recall problems, or for that matter slips of the tongue (or pen), can be accurately identified as "Freudian slips." There are many cognitive imperfections of which very few may be psychologically motivated in the way that Freud supposed.

As shown in Table 2.1, dream consciousness has very different rules for memory access and storage from waking consciousness. It is rare, in dreaming, to seek a memory, as in the increasingly common waking experience that I have (due to aging) of trying to remember the name of a place or a person. This failure to recollect voluntarily could be due to the loss of self-reflective awareness that I will discuss in the next section. And it is certainly related to the occurrence of dream plot discontinuity and incongruity as well as to my failure to recognize these gross peculiarities when I dream. This has long been recognized by dream interpreters who make of it an active process – a psychologically motivated need to forget–but their emphasis is upon poor dream recall following awakening and not on the striking amnesia that occurs *within* the dream itself – like my inability to realize that I can not possibly be riding a bike in my recent dreams because I lost that ability 7 years ago!

My colleague, Ursula Voss, a psychologist who works in Bonn, Germany, has been studying the dreams of persons who are blind, deaf, mute, and paraplegic from birth (Voss et al. 2011). Such subjects do not dream of themselves as handicapped nor do they experience handicap in their dreams. It is as if their protoconscious brain saw, spoke, heard, and moved in their dreams.

In another recent dream, I embraced my lifelong friend, the pianist Rial Ogden, who was playing very sweetly for me. Suddenly I realized sadly that Rial had died about a year ago. I said "Rial, you can't leave me," as I hugged him tightly. In this case I became partially lucid but I still didn't know I had been dreaming until I woke up.

Why, then, are certain dream plots formed and not others? My Rial Ogden-pianist dream, is not hard to understand. It was triggered by a longing for Rial. But what about tree trimming in London? That one is not so easy. In fact, Magdalena Fosse and others in our group found that dreamers were not able to identify *any* specific memory source for about 20 % of their dream plot items (Fosse et al. 2003). So, contrary to the popular idea that dreams replay memories, we need to consider the possibility that there is often no discernible memory order in dream plot selection and construction.

This idea is repugnant because it suggests that chance plays a larger part in the shaping of our dream lives than we would like to admit. But the idea becomes more attractive when we recognize that a random process guarantees a more thorough check on the contents of memory than an overly determined mechanism would allow. In other words, in exchange for the comfort of interpretive constraint we gain considerable cognitive freedom (and even creativity). The problem is that we have to abandon our interpretive conceits to recognize it!

Many people prefer the security of just-so stories to the anxiety of open-field uncertainty. Like Freud, they are Newtonian mechanists rather than Einsteinian relativists or, God forbid, those quantum physicists like Niels Bohr (1934), who posits that two contrary states can simultaneously co-exist!

Speaking of security, my London tree trimming dream makes it clear that my upper brain still knows how to ride a bike. That skill could be reinforced during sleep even if it is no longer useful to me in waking. As for the tree trimmer, that too I still know how to use. In fact, when awake I even know exactly where in my Vermont barn it is stored even though I will probably never use it again (in waking, that is!).

There is more abundant evidence to support the idea that at least procedural memory benefits from sleep (see Fig. 4.4). This fits with the finding of formal analysis regarding the ubiquitous movement in dream plots, a robust formal feature never previously noticed by interpreters of dreaming. McCarley and Hoffman pointed out that every sentence of every REM sleep dream report contained an action verb (McCarley and Hoffman 1981). Dream consciousness is much more concerned with doing than it is with reflecting. We will further emphasize this point when we discuss dream thinking.

The idea that is lurking here is that the apparent down-time of sleep is an illusion. Even if there are no outward signs of movement in REM sleep, motor programs in the upper brain are churning away. Internal networks can be checked while we sleep peacefully and have only slight recollections of this iterative process when we wake up. Developmental theories are very compatible with this sort of insight. Not only are we able to practice our skills while we sleep, but we might be able to produce and improve those skills in a safe behavioral vacuum.

This is indirect but strong evidence for the protoconsciousness hypothesis. Not only is the dreaming brain prepared to act in waking but it is prepared to act in an effective way. This is more than scenario framing, it is action enhancing! I don't need to be conscious of how to ride a bike. My brain just knows how to do so because it has run my bike riding programs while I was asleep!

8. *Dream Emotion*

My London tree trimming dream is not highly emotional. In that dream, I did have some anxiety related to my orientational uncertainty but my wake state is mild compared to that of many of my dreams. I am not particularly elated by my dream obligations nor am I as miffed as might be expected by having pedaled halfway across London to trim that puny tree. I was a bit surprised to find my prospective employer and his daughter dressed in eighteenth century garb but surprise, that cousin of emotion, was quickly squashed by the congruity of the costumes with the eighteenth century architecture that I had noted. This typical *ad hoc* rationalization shows that my dream thinking is motivated but uncritical.

Even when dreaming is much more emotional than my London gardening adventure was, dream reports do not include many descriptions of feelings unless subjects are asked to give them. Dreamers are so busy describing the curious details like riding the bike in traffic with a 6 ft. long tree trimmer across the handlebars, the Georgian neighborhood and its denizens, and the puny little tree in the garden, that they forget to mention the usual dream feelings of anxiety, elation or anger unless specifically instructed to do so.

Insofar as I had any emotion in my tree trimming dream, that emotion was entirely consonant with the bizarre plot. Compared to waking, I would have felt much more anxious about cycling in London traffic with an unwieldy tool across my handlebars than I did in my dreams. And I really should have been much more put out to be summoned to trim a shrub that was so scrawny that I myself would have cut it down. As mentioned before, I would certainly have fretted over how I was to be compensated for this absurd inconvenience.

Come to think of it, my emotional state in the tree trimming dream was surprisingly flat. I have something like the '*belle indifference*' of Jean-Martin Charcot and Pierre Janet's hysterical patients at the Salpêtrière Hospital in Paris. Normally emotion, especially if elicited with affirmative probes, is quite prominent in reports of dream consciousness. As suggested above, the most common dream emotions are anxiety, fear, elation, and anger. Strikingly under-expressed are what we might call such social emotions as shame, guilt, and sadness. I had none of these emotions as I cycled across London.

As a rule dream emotion (as in waking) is entirely consonant with dream plot details. In other words, emotion is tightly associated with what is going on in the dream. If I were thoroughly lost, as I so often am in my dreams, I should be extremely anxious. If I were headed for a more exciting dream interlude, like a romantic encounter, I might well be giddily elated. A hostile encounter with my prospective employer would have enraged me. This consonance between dream emotion and dream content is what I call emotional salience. In that sense, dreaming and waking consciousness are quite similar even though the emotional range of dreaming is constricted.

Although not yet tested by systematic experiment, it is my impression that dream consciousness is more fraught with emotion than is waking consciousness. When I consider the long hours that I spend awake alone, reading, writing, and thinking, I would estimate my conscious, waking emotional experience to be very near zero. If the telephone rings, I may have a violent but very brief startle response which causes me to shift my attention from the private world of my work to the public world around me. There may also be a faintly perceived level of what I call existential anxiety but it is nothing like my intensely anxious dreams. Of course, you might say, a waking control for dream emotion must be social, not private. The content of dreaming is social rather than individual.

Leaving aside the important formal contextual difference between wake and dream consciousness, it is tempting to argue that I was mostly alone in my London bicycle dream and therefore had little social occasion for emotion. But I am often alone in my dreams and often very anxious in association especially with extreme disorientation. So it is not as simple as the social context argument would suggest. There is no substitute for an empirical test. However challenging an assessment of wake-state emotion may be, experimental investigation must be undertaken.

As already emphasized, dream consciousness is commonly emotional with anxiety, elation, and anger leading the parade. But the emotional element is qualitatively and quantitatively variable in ways that we don't yet fully understand. While many dream emotions make sense from a contextual point of view, there are notable exceptions suggesting that even in this cardinal area there are discontinuities as well as tight links. The emotional-cognitive system usually, but not always, plays by a set of strict associative rules.

These and other cognitive features of dreaming are so distinctively different from those of waking as to suggest a major reorganization of underlying brain activity. The details of that reorganization will be the subject of Lecture II. As early as 1953, it was recognized that brain physiology did change dramatically in sleep (Aserinsky and Kleitman 1953). The superficial signs of these profound changes are summarized in Fig. 2.2.

9. *Dream Thinking*

Compared to my own waking life, I pay far less attentional energy to thinking when I am dreaming. Even though I am an academic professional and therefore paid to think in waking, I cannot use that fact to explain the extreme paucity of thinking during my dreams. As I cycled through London with a tree trimmer across the handlebars of my bike, I never once questioned my activity or my goal. I was in the thrall of Allan Rechtschaffen's single mindedness. I was dream-bicycling, but I was not thinking.

I did not have the usual background concerns that characterize my waking consciousness. When I am going out to a teaching assignment or making a house call for medical reasons, I am always asking myself: Am I on the right route? What street is this? Where is my map? Whom do I expect to meet? What is our contract? What will I say? These background cognitive elements are typical of what is called waking fantasy. Sometimes fantasy is magically positive (I will succeed certainly). But more often than not, it is banal, commonplace or even negative in tone (defeat is inevitable). This feature has led me to conclude that much so-called fantasy is in fact fictitious behavioral rehearsal and predictive preparation for social interaction. This gives waking consciousness one of its greatest powers: autocriticism, an important part of self-reflective awareness. This is a feature of what we call secondary consciousness. It is not present in dreams so it must be specific to waking.

Autocriticism is related to theory of mind insofar as it is an awareness of one's social impact upon another person. As we speed on bikes or in cars to a social rendezvous, our foreground consciousness guides the vehicle to its destination while we play out social scenarios in our mind. Being able to imagine a variety of scenarios is to help us prepare our response to them. Of course real life scenarios may defy prediction. Then we are surprised and have to fly by the seat of our pants and react instinctively.

In dream consciousness we have no background awareness, hence no alter ego, no self-reflective awareness, no monitor, and no working memory. I don't

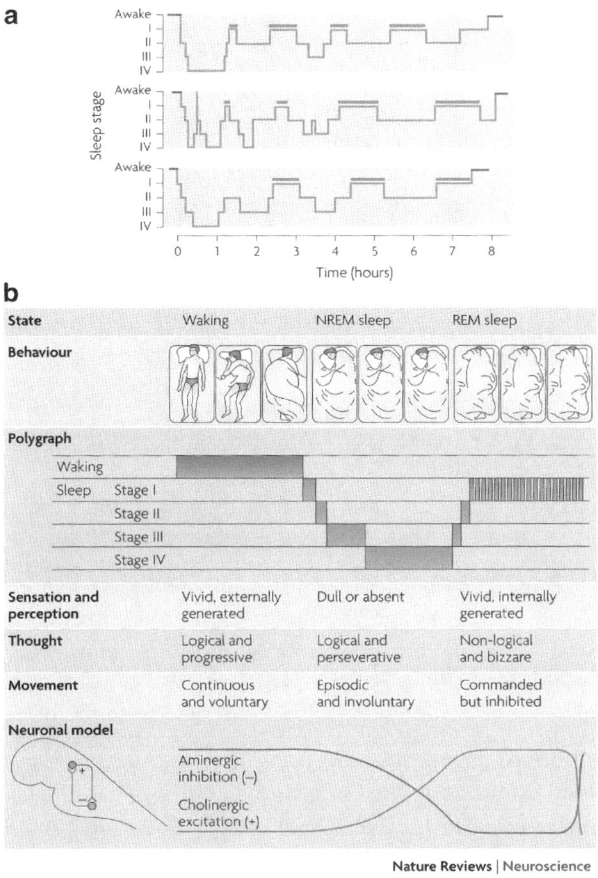

Fig. 2.2 Sleep science fundamentals
Standard sleep laboratory measurements reveal 90–100 min cycles of rapid eye movement (*REM*) and non-rapid eye movement (*NREM*) sleep (see the figure, part **a**, which shows cycles for three subjects; the *blue lines* indicate periods of REM sleep). Reports of dreaming are most common from sleep onset stage I (when dreams tend to be fragmentary and unsustained), late-night stage II (when dreams tend to be thought-like) and stage I REM (when they tend to be long, vividly hallucinatory and bizarre). All of the deep phases of sleep (III and IV) occur in the first half of the night, whereas lighter stages of sleep (stages I and II) predominate in the second half of the night. Regardless of time of night, reports of dreaming are longest and most bizarre following awakenings from stage I REM.
The states of waking and sleep have behavioral, polygraphic and psychological manifestations (see the figure, part **b**), which seem to be orchestrated by a control system in the pontine brainstem. In the figure, the neuronal model of the clock that programs these states is depicted as the reciprocal interconnection between aminergic inhibitory neurons and cholinergic excitatory neurons. The activity levels of these two groups of neurons are also reciprocal: the activity of aminergic cells is highest during waking, declines during NREM sleep and is lowest during REM sleep, whereas the activity of cholinergic cells shows the reverse pattern. Changes in state and phase take place whenever the two activity curves cross; these are also the times when major postural shifts occur. The motor immobility during sleep depends on two different mechanisms: disfacilitation during stages I–IV of NREM sleep and inhibition of motor systems during REM sleep. The motor inhibition during REM sleep prevents motor commands from being carried out, so that we do not act out our dreams. (Part a is unpublished data of Snyder and Hobson) (Part **b** of the figure is reproduced, with permission, © (1983) J. Allan Hobson. Parts **a** and **b** reproduced in Hobson 2009) (Color figure online)

ask myself if it is wise to ride by my bike in London traffic. I don't ask why I should be carrying a tree trimmer. It is not so much that my thinking is illogical as it is that I am nearly literally thoughtless when I dream. My waking behavior would suffer enormously if it were guided by dream consciousness.

Roar Fosse quantified this dream thoughtlessness when he measured explicit references to thinking in the 3,000 reports he obtained from our 'grand mentation' study of 16 people as they woke, worked, made love and slept, over a 2 week period (Fosse et al. 2001). In contrast to quiet or more active waking those subjects showed a twofold drop in explicit reference to thinking while they were dreaming (in REM sleep compared to quiet or active waking). These results are illustrated in Fig. 2.3.

Recently David Kahn and I (Kahn and Hobson 2005) have returned to this problem. We asked two questions:

(1) Was Roar Fosse correct in his observation that thinking declined markedly in REM sleep?
(2) Was such thinking as was present logically sound, or was it logically flawed? (a question that Roar Fosse didn't ask).

To answer these two questions we began by studying 40 dreams recorded by me over 2 years ago (like the London tree trimming dream).

Compared to waking, the dream reports showed the predicted decrease in the initiation of thinking. While some dream logic was as inferentially sound as it is in waking, some was inferentially unsound. One example of each follows:

(1) Sound Reasoning:

  I infer from seeing the eighteenth century houses that I must be in the right part of the town. (Given the premise of my imagined destination, this inference is logical).

(2) Unsound Reasoning:

  I infer from the dress of my employer that this must be a pageant day for the neighborhood. (This inference is superficially logical but is a far-fetched *ad hoc* explanation of an unexpected dream observation).

  Most striking of all was the absence of any inferential operation by my mind on even outlandishly bizarre dream plot items. For example there was never any reasonable question about why the tree trimmer was being carried. The danger of the traffic and the ultimate redundancy of the tool simply did not register in my dream consciousness.

  At this point you might well ask: 'Why all this fuss about logical thinking in dreams? We have always known that dreams were not logical.' But, as shown above, some dream thinking is present and, when it is present, the thinking may be logically sound. Research psychologists have therefore questioned the faulty logic hypothesis as part of their tendency to equate waking and dreaming consciousness. As silly as that may seem to us, it is important to take peer opinions seriously and to determine that the observed distinctions are due to methodological, not substantial differences between

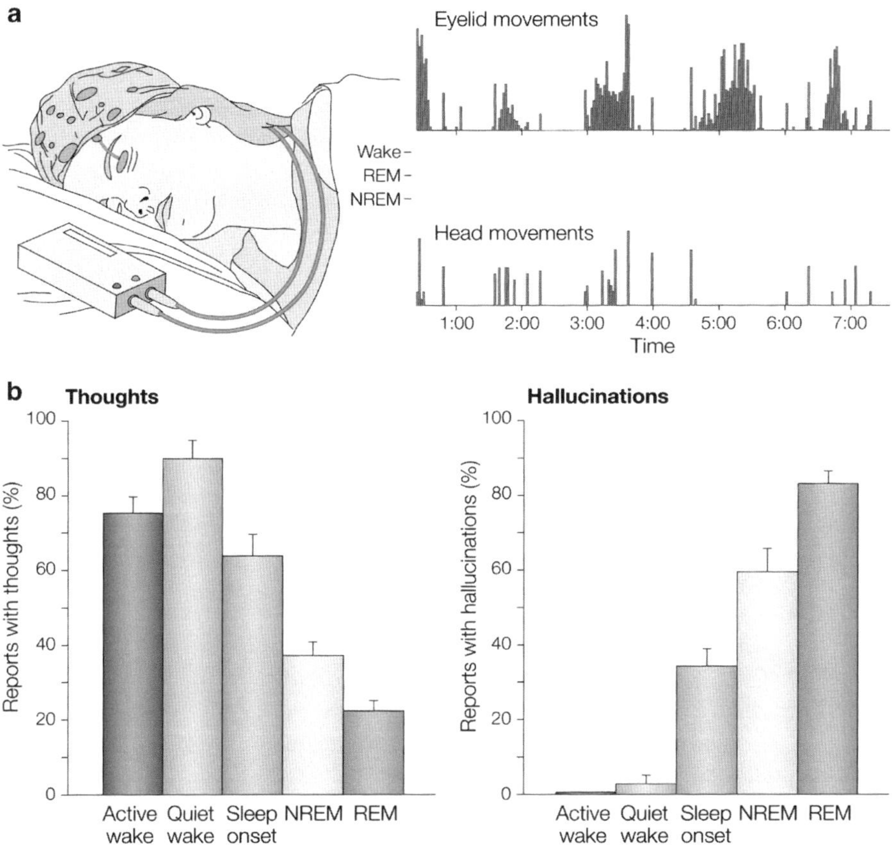

Fig. 2.3 Reciprocal state-related changes in thoughts and hallucinations measured using the Nightcap system. (**a**) Central arousal accompanying the activated states of rapid eye movement (*REM*) sleep and waking can be measured using the "Nightcap"–a simple ambulatory monitor. The Nightcap is a two-channel recording device that distinguishes waking, REM sleep and non-REM (*NREM*) sleep. One channel of the Nightcap monitors eye movement and the other monitors body movements. The Nightcap eyelid-movement readout is thought to reflect activity in portions of the brainstem oculomotor nucleus that innervate the eyelid and are adjacent to portions of the medial brainstem ascending reticular system, the activity of which, in turn, generates forebrain activation. (**b**) Decline in directed thought and reciprocal increase in hallucinations during progression from active waking through sleep onset and NREM sleep to REM sleep. The values of thoughts and hallucinations are reciprocal for the five states represented on the abscissas of the two graphs. Thoughts and hallucinatory imagery would appear to be mutually antagonistic and, in the extreme, incompatible with each other (Reproduced with permission from Hobson and Pace-Schott 2002. Original modified from Fosse et al. 2001)

studies. Because thinking is so rare, our dream consciousness is very different from our consciousness in waking. We need to establish that fact beyond the shadow of a doubt before we ask why.

10. *Insight and Judgment*

At no time during my London bicycle dream did I entertain any doubt that I was awake. And yet I was asleep. My insight regarding my true state was therefore entirely non-existent. I was completely fooled. I should have asked myself, even if I was fooled, about my true state. "Why should I be riding a bike in heavy London traffic, and with a tree trimmer across the bike's handle bars?" I am a physician by trade, not a tree trimmer and while I do sometimes visit London, I never ride a bike there. Not even before I had my stroke did I ever ride a bike in London. The knowledge base necessary to achieving this insight is simply unavailable to me when I am dreaming. It is walled off from conscious access, like certain aspects of mental activity that are made unavailable to waking consciousness by a hypnotist.

Without insight, my judgment is also poor. I fail to recognize the danger to which I exposed myself by riding a bike in London traffic. I admit that I used to pedal dangerously around Boston, but that was 50 years ago and I knew at the time that my judgment was poor. No such awareness characterized my dream consciousness. In my right senses, I would never expose myself to the ridicule that might be heaped upon an itinerant tree trimmer. And had I beheld a girl and her father dressed up as eighteenth century gentry, I would have laughed and gone straight home. The scrawny tree in their yard would have helped me to turn away and to chuckle at my own foolishness.

11. *Abstraction* versus *Concreteness*

One of the most powerful and delightful attributes of waking consciousness is our capacity to think abstractly. Abstraction is the essence of art and science. When we see and experience an object, a person, an event, we are not only able to classify and characterize that particular stimulus but also to generalize from it. We thus recognize that our perception is only superficial and that a deeper meaning lies beneath the surface of each observation.

We say that a sign is also a symbol so that one thing can stand for another. Dream interpreters would not accept my bicycle as only a bicycle, but perhaps, a social vehicle from the wrong to the right part of town; my tree trimming tool might also be seen as a practical ticket for this social voyage. Even though I don't need it, and didn't use it, it is there defending me against the charge of decadence. For the died-in-the-wool psychoanalyst, a tree trimmer is a more provocative instrument because it culls and cuts branches up to an inch thick. It would be wise to keep fingers and other long thin body parts out of its reach!

I have none of these ideas when I dream. It is only when I wake up and consider my dream that my waking consciousness makes these suggestions. So it could be that dreaming is phenomenologically concrete but that dreams are inevitably abstract. Hence, they invite interpretation. But how can we have more than literary satisfaction from any such interpretation? I look forward to hearing your answers to this question.

By contrast, the rigidification of the mind by whatever changes occur in sleep may be specifiable. At this point we could say: 'O.K., Dreams are interpretable in the same way that any product of human consciousness is interpretable.' But we should certainly not assume that such abstraction as is inspired by even concrete mental products was put there to deceive us.

On that reading I am not supposed to know I am a social climber or a man who is determined to be respected as much for his manual skills as for his intellect. The 18th century has always had an allure for me, so why not stay and enjoy the party. Stash the bike and chuck the tree trimmer. If you are in a time warp, go with it. Put on some frilly clothes and join the party. I might even hear the following poem, written in 1724, read for my pleasure by its frock-coated author disguised as Jonathan Swift (1823):

Those Dreams that on the silent night intrude,
And with false flitting shades our minds delude,
Jove never sends us downward from the skies,
Nor can they from infernal mansions rise,
But are all mere productions of the brain,
And fools consult interpreters in vain.

Sleep and Dreaming.
Jay Vogelsong has created this image in the manner of Giorgio de Chirico to create a sense of dream disorientation.

2.3 Lucid Dreaming: Curing the Psychosis of Dreaming by Autosuggestion

Just as psychotic patients can be trained, through psychotherapy, to recognize their hallucinations and delusions as fabrications, so can normal people learn to recognize their dream hallucinations and delusions for the cognitive miscreants that they are. This rare and evanescent state is called lucid dreaming. In the sleep lab, a pre-arranged sequence of voluntary eye movements indicates a subject's awareness that he is dreaming, while he is dreaming. This dream double-mindedness occurs spontaneously and is quite common in children of 9–10 years of age after which it declines in spontaneous incidence. But it can be inculcated at least until the age of 40, by simple pre-sleep autosuggestion. After age 40, lucidity declines with many other delights of life.

With a notebook and pen on the bedside table as a task reminder and recorder, you can tell yourself just before going to sleep that you are going to look for and recognize bizarreness as a sign that you are dreaming. As soon as you notice discontinuity and incongruity in your mental content you will know that you are dreaming. That will allow enough of you to wake up enough to watch the dream evolve. You may even want to intervene, change the plot, and enact pleasurable behaviors like flying and love-making.

Ursula Voss and her colleagues have recently demonstrated that young subjects, who are trained to signal their dream lucidity via voluntary eye movements, have a significant increase in the 40 Hz EEG activity of their frontal lobes. They also show an increase in the EEG coherence of their frontal and occipital lobes. These findings suggest that lucid dreaming is more like waking than non-lucid dreaming. Consciousness clearly changes its qualitative as well as its quantitative character when underlying brain activation is globally or regionally altered (Voss et al. 2009). These results are illustrated in Fig. 4.3.

Lucid dreaming is as evanescent as it is rare. Subjects note that they may lose their power to observe as they are pulled back down into the dream. Or they may wake up altogether. In that case they are no longer dreaming but can have recall. And having assured recall by awakening it may be possible to return, directly, to the same dream plot that the dreamer has temporarily exited! As if we didn't know it already, this proves beyond the shadow of a doubt that dreaming is as much a psychodynamic process as it is an organic one. The argument is thus in favor of an and/or rather than an either/or approach to the brain mind. The brain-mind, it would seem, can change itself.

I will say more about brain physiology in the next lecture. My main point here is to emphasize the importance and the richness of relatively simple approaches to the psychology of dream consciousness. A key to success with this approach is to address formal rather than content analytic aspects of the dreams. Despite some similarities, dream consciousness is radically different from waking consciousness. How are we to explain the similarities and differences and what functional sense can we make of them? The answer to the last question could be quite different from that suggested by content analysts like Sigmund Freud. My answer is that REM sleep dreams are evidence for a protoconsciousness process that indicates the brain-mind preparation for its most glorious achievement, waking consciousness.

References

Aserinsky, E., & Kleitman, N. (1953). Regularly occurring periods of ocular motility and concomitant phenomena during sleep. *Science, 118*, 361–375.

Bohr, N. (1934). *Atomic theory and the description of nature*. Cambridge: Cambridge University Press.

Fosse, R., Stickgold, R., & Hobson, J. A. (2001). Brain-mind states: Reciprocal variation in thoughts and hallucinations. *Psychological Science, 12*(1), 30–36.

Fosse, R., Stickgold, R., & Hobson, J. A. (2002). Emotional experience during rapid-eye-movement sleep in narcolepsy. *Sleep, 25*(7), 724–732.

Fosse, M. J., Fosse, R., Hobson, J. A., & Stickgold, R. J. (2003). Dreaming and episodic memory: A functional dissociation? *Journal of Cognitive Neuroscience, 15*(1), 1–9.

Hobson, J. A. (1983). *Sleep: order and disorder* (Behavioral biology in medicine series). South Norwalk: Meducation.

Hobson, J. A. (2009). REM sleep and dreaming: Towards a theory of protoconsciousness. *Nature Reviews Neuroscience, 10*, 803–814.

Hobson, J. A., & Brazier, M. A. B. (Eds.). (1980). *The reticular formation revisited* (Vol. 6). New York: Raven.

Hobson, J. A., & Kahn, D. (2007). Dream content: Individual and generic aspects. *Consciousness and Cognition, 16*(4), 850–858.

Hobson, J. A., & Pace-Schott, E. F. (2002). The cognitive neuroscience of sleep: Neuronal systems, consciousness and learning. *Nature Reviews Neuroscience, 3*(9), 679–693.

Hobson, J. A., Hoffman, S. A., Helfand, R., & Kostner, D. (1987). Dream bizarreness and the activation-synthesis hypothesis. *Human Neurobiology, 6*(3), 157–164.

Hobson, J. A., Pace-Schott, E. F., & Stickgold, R. (2000). Dreaming and the brain: Toward a cognitive neuroscience of conscious states. *Behavioral and Brain Sciences, 23*(6), 793–842.

Hobson, J. A., Sangsanguan, S., Arantes, H., & Kahn, D. (2011). Dream logic: The inferential reasoning paradigm. *Dreaming, 21*(1), 1–15.

Kahn, D., & Hobson, J. A. (2005). State-dependent thinking: A comparison of waking and dreaming thought. *Consciousness and Cognition, 14*(3), 429–438.

Kahn, D., Stickgold, R., Pace-Schott, E. F., & Hobson, J. A. (2000). Dreaming and waking consciousness: A character recognition study. *Journal of Sleep Research, 9*(4), 317–325.

Kahn, D., Pace-Schott, E., & Hobson, J. A. (2002). Emotion and cognition: Feeling and character identification in dreaming. *Consciousness and Cognition, 11*(1), 34–50.

McCarley, R. W. (1982). REM sleep and depression: Common neurobiological control mechanisms. *The American Journal of Psychiatry, 139*(5), 565–570.

McCarley, R. W., & Hoffman, E. (1981). REM sleep dreams and the activation-synthesis hypothesis. *The American Journal of Psychiatry, 138*(7), 904–912.

Merritt, J. M., Stickgold, R. A., Pace-Schott, E. F., Williams, J., & Hobson, J. A. (1994). Emotion profiles in the dreams of young adult men and women. *Consciousness and Cognition, 3*, 46–60.

Pace-Schott, E. F., & Hobson, J. A. (2002). The neurobiology of sleep: Genetics, cellular physiology and subcortical networks. *Nature Reviews Neuroscience, 3*, 591–605.

Porte, H., & Hobson, J. A. (1996). Physical motion in dreams: One measure of three theories. *Journal of Abnormal Psychology, 105*, 329–335.

Rechtschaffen, A. (1978). The single-mindedness and isolation of dreams. *Sleep, 1*, 97–109.

Resnick, J., Stickgold, R., Pace-Schott, E., Williams, J., & Hobson, J. A. (1994). Self-representation and bizarreness in children's dreams. *Consciousness and Cognition, 3*, 30–45.
Scarone, S., Manzone, M. L., Gambini, O., Kantzas, I., Limosani, I., D'Agostino, A., & Hobson, J. A. (2008). The dream as a model for psychosis: An experimental approach using bizarreness as a cognitive marker. *Schizophrenia Bulletin, 34*(3), 515–522.
Stickgold, R., Pace-Schott, E., & Hobson, J. A. (1994a). A new paradigm for dream research: Mentation reports following spontaneous arousal from REM and NREM sleep recorded in a home setting. *Consciousness and Cognition, 3*, 16–29.
Stickgold, R., Rittenhouse, C., & Hobson, J. A. (1994b). Dream splicing: A new technique for assessing thematic coherence in subjective reports of mental activity. *Consciousness and Cognition, 3*, 114–128.
Stickgold, R., James, L., & Hobson, J. A. (2000). Visual discrimination learning requires sleep after training. *Nature Neuroscience, 3*(12), 1237–1238.
Sutton, J., Rittenhouse, C., Pace-Schott, E., Stickgold, R., & Hobson, J. A. (1994a). A new approach to dream bizarreness: Graphing continuity and discontinuity of visual attention in narrative reports. *Consciousness and Cognition, 3*, 61–88.
Sutton, J., Rittenhouse, C., Pace-Schott, E., Merritt, J., Stickgold, R., & Hobson, J. A. (1994b). Emotion and visual imagery in dream reports: A narrative graphing approach. *Consciousness and Cognition, 3*, 89–99.
Swift, J. (1823). *The select works of Jonathan Swift, D.D.* (Vol. 5). London: Published for Hector McLean.
Voss, U., Holzmann, R., Tuin, I., & Hobson, J. A. (2009). Lucid dreaming: A state of consciousness with features of both waking and non-lucid dreaming. *Sleep, 32*(9), 1191–1200.
Voss, U., Tuin, I., Schermelleh-Engel, K., & Hobson, J. A. (2011). Waking and dreaming: Related but structurally independent. Dream reports of congenitally paraplegic and deaf-mute persons. *Consciousness and Cognition, 20*(3), 673–687.
Walker, M. P., Brakefield, T., Morgan, A., Hobson, J. A., & Stickgold, R. (2002). Practice with sleep makes perfect: Sleep-dependent motor skill learning. *Neuron, 35*(1), 205–211.
Walker, M. P., Brakefield, T., Hobson, J. A., & Stickgold, R. (2003). Dissociable stages of human memory consolidation and reconsolidation. *Nature, 425*(6958), 616–620.
Williams, J., Merritt, J., Rittenhouse, C., & Hobson, J. A. (1992). Bizarreness in dreams and fantasies: Implications for the activation-synthesis hypothesis. *Consciousness and Cognition, 1*, 172–185.

Chapter 3
Lecture II: Physiology

J. Allan Hobson

To Dream, Perchance to Think

In the beginning was the dream-state brain.
For, long before the word (and with it, God)
Evolved on earth to stake out man's domain
There first emerged a state in which brains nod.
From out the neural populations urged
By amines or by cholinergic sway
A climate formed wherein these two diverged
To shape contrasting dreams of night and day:
The first as atlas of the world outside
On which to read (and write) our way; the next
As what we find out there (once it's transcribed
In that same ink in which is writ both texts).
Hence, input from outside and in were wrought
Commingled as one mind, begetting thought.

–David Borodin

In order to explain the myriad psychological distinctions between dreaming and waking that were detailed in Lecture I, it is useful to analyze them a bit so as to know what questions to put to physiology. Then it is helpful to organize the equally bewildering panoply of physiological findings according to a few basic principles. The brain anatomy which is most germane to the following discussion is illustrated in Fig. 3.1.

The first issue is how to account for the occurrence of dreaming in sleep. The general physiological principle is that brain activation is greatest in REM sleep when the most

J.A. Hobson (✉)
Department of Psychiatry, Harvard Medical School, Boston, MA, USA
e-mail: allan_hobson@hms.harvard.edu

N. Tranquillo (ed.), *Dream Consciousness*, Vienna Circle Institute Library 3,
DOI 10.1007/978-3-319-07296-8_3, © Springer International Publishing Switzerland 2014

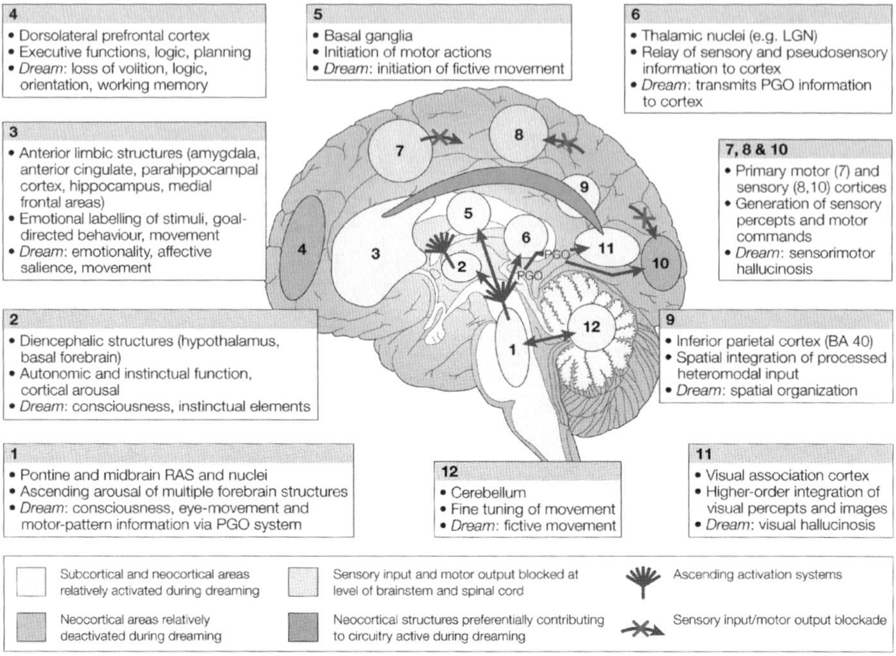

Fig. 3.1 Forebrain processes in normal dreaming – an integration of neurophysiological, neuropsychological and neuroimaging data. Regions 1 and 2, ascending arousal systems; 3, subcortical and cortical limbic and paralimbic structures; 4, dorsolateral prefrontal executive association cortex; 5, motor initiation and control centers; 6, thalamocortical relay centers and thalamic subcortical circuitry; 7, primary motor cortex; 8, primary sensory cortex; 9, inferior parietal lobe; 10, primary visual cortex; 11, visual association cortex; 12, cerebellum. *BA* Brodman area, *LGN* lateral geniculate nucleus, *PGO* ponto-geniculo-occipital, *RAS* reticular activating system (Reproduced, with permission, from Hobson and Pace-Schott 2002)

sustained and intense dreaming occurs (see again, Fig. 2.2). Activation (A) is also considerable in NREM sleep which is capable of sustaining some mental activity akin to REM sleep dreaming especially in the second half of the night. Please note that at its most deactivated nadir, in stage IV of NREM sleep, the brain is still 50–80 % activated. The brain never turns off. It continues to work all the time that we sleep. No wonder we are sometimes quite conscious in our sleep. No wonder that we often dream.

But how is it that we do not wake up when our brain is activated, especially in REM sleep when dreaming is most intense? The answer is that in REM sleep the brain closes its input and output gates so that it is effectively offline with respect to external sensory stimulation. The brain changes its information source (I) from predominantly external to almost exclusively internal. And, due to inhibition of the spinal cord, it does not enact its own motor commands. A notable exception to the quelling of motor output is the occurrence of rapid eye movements that give REM its name but which do not wake us up. In other words, brain actively inhibits the external stimuli which are an important shaper of waking consciousness; it also actively inhibits the motor output which accompanies waking behavior.

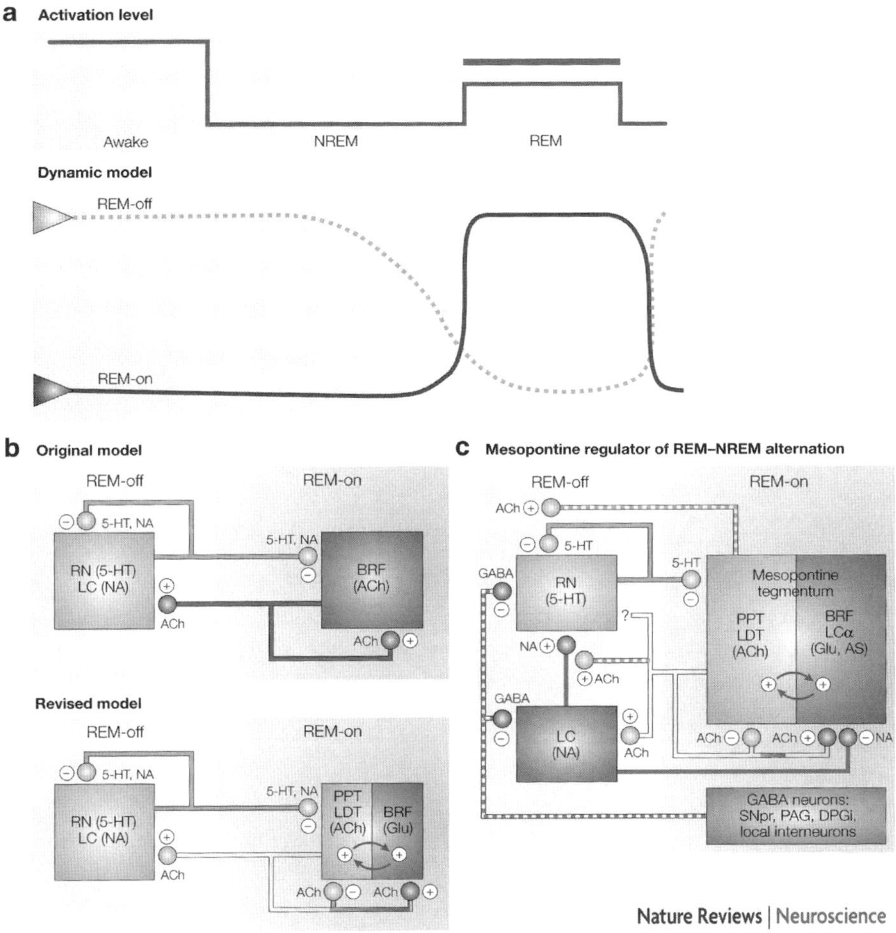

Fig. 3.2 The reciprocal-interaction model of REM-NREM sleep alternation. (**a**) Activation model showing the alternation of rapid eye movement (*REM, red bar*) and non-REM (*NREM*) sleep states, and the underlying dynamic model of alternating activity of REM-on and REM-off cells in pontine nuclei. Although the form of the original schematics that depicted the reciprocal-interaction model (**b**) appears to indicate that the release of acetylcholine (*ACh*) is low during waking and increases only with REM, it should be noted that ACh is a neurotransmitter of the activated brain and, as such, is increased equally during waking and REM sleep. (**b**) The original reciprocal-interaction model of NREM-REM alternation and the subsequent revision of this model to reflect the findings of self-inhibitory cholinergic autoreceptors in the mesopontine cholinergic nuclei, and excitatory interactions between mesopontine cholinergic and non-cholinergic neurons. *BRF* brainstem reticular formation, *Glu* glutamate, *5-HT* 5-hydroxytryptamine (serotonin), *LC* locus coeruleus, *LDT* laterodorsal tegmental nucleus, *NA* noradrenaline, *PPT* pedunculopontine tegmental nucleus, *RN* raphé nuclei. (**c**) Further synaptic details of the revised reciprocal-interaction model shown in (**b**). These additional synaptic details can be superimposed on the revised model without altering the basic effects of aminergic and cholinergic influences on the REM sleep cycle that are depicted in (**a**) and (**b**). Excitatory cholinergic-non-cholinergic interactions involving ACh and the excitatory amino-acid transmitters enhance the firing of REM-on

Where, then, do dream stimuli come from? They must come from the brain itself and so they do. One important internal stimulus source is the eye movement system, which sends information about its action from the brainstem to other parts of the brain including the thalamus and cortex which most scientists assume to be essential to consciousness. Since there is no sensory input, as prevail in waking, this internally generated data is read as if it were coming from the outside world. The brain's model of the outside world is thus both activated and stimulated. As far as the brain knows, it is awake because it is activated and stimulated.

But there is more to the story than this. The kind of consciousness that the brain assumes in dreaming is quite different from that of waking. Contributing to this differentiation is the major biochemical difference in brain activation that results from the change in brain chemical modulation (M). Specifically, there is a dramatic decline in the secretion of the aminergic neuromodulators serotonin, norepinephrine, and histamine in REM sleep (See Fig. 3.2). While these systems turn off in sleep, the GABA, acetylcholine, and dopamine systems continue to function, if not entirely normally, at least significantly. In waking, the brain is activated *and* aminergically modulated. In REM sleep, the brain is activated but aminergically de-modulated. With respect to chemical modulation, the two brain activated states, waking and REM sleep, could not be more different from each other.

The activation-synthesis hypothesis attributes dreaming to brain activation in sleep. The theory states that dreaming is the subjective awareness of brain activation in sleep and that the sleep activation of the brain results in the synthesis of conscious elements (e.g. emotion, perception, and thinking). It further suggests that the natural activation process stimulates the brain in such a way as to create a wake-like state which is offline to the outside world. The self-activated, offline brain considers itself to be awake because it is deprived of its working memory. It also lacks self-reflective awareness, insight and judgment. These psychological functions depend upon the neuromodulators that are active in waking but shut down in sleep. We will consider regional effects when we have taken up activation (A) in detail and, again, when we describe the changes in information source (I) and modulation (M).

Once I have reviewed the evidence for A, I, and M, we will combine the measures of these functions to create a three dimensional model of the brain mind. In this new model called AIM, time is the fourth dimension (See Fig. 3.5). The cycle of brain-mind states that occurs in waking and sleep can be displayed as the successive values of AIM.

Fig. 3.2 (continued) cells, whereas inhibitory noradrenergic, serotonergic, and autoreceptor cholinergic interactions suppress REM-on cells. Cholinergic effects on aminergic neurons are excitatory, as proposed in the original reciprocal-interaction model, and might also operate by presynaptic influences on noradrenergic-serotonergic as well as serotonergic-serotonergic circuits. GABA (γ-aminobutyric acid) and other neurotransmitters, such as nitric oxide and neuropeptides, might contribute to the modulation of these interactions. *AS* aspartate, *DPGi* dorsal paragigantocellular nucleus, *LCα* peri-locus coeruleus-α, *PAG* periaqueductal grey area, *SNpr* substantia nigra pars reticularis (Reproduced, with permission, from Pace-Schott and Hobson 2002) (Color figure online)

3.1 Activation (Factor A)

Until the discovery of the reticular activation system by Giuseppe Moruzzi and Horace Magoun (Moruzzi and Magoun 1949) in 1949, most scientists followed Frederic Bremer, Charles Sherrington and Ivan Pavlov in assuming that brain activation depended upon sensory input (Bremer 1935; Sherrington 1906). For them, sleep was a simple deactivation process related to what came to be called the deafferentation hypothesis: the brain turned off owing to the diminution of sensory input.

This deafferentation theory died hard. First formulated by Bremer in the 1930s, it was unequivocally done in by Aserinsky and Kleitman's discovery, in 1953, of automatic, regularly periodic, and spontaneous brain activation in sleep. The REM sleep activation, discovered by Aserinsky and Kleitman in Chicago (Aserinsky and Kleitman 1953), was not only wholly independent of sensory input, it was also the physical substrate of hallucinoid dreaming, a psychological state with many wake-like features but with its own distinctive attributes which cried out for psychophysiological interpretation. The neuronal basis of activation and deactivation was established by the late Mircea Steriade (Steriade 1981) whose brilliant work is illustrated in Fig. 3.3.

Modern research on the psychophysiological basis of consciousness can be said to have begun in 1953, the same year that James Watson and Francis Crick (Watson and Crick 1953), working at Cambridge in England, deduced the molecular structure of DNA. An integration of Moruzzi and Magoun's reticular activation concept and Aserinsky and Kleitman's REM sleep discovery was made by the French neurosurgeon, Michel Jouvet, working in Lyon, France. Jouvet performed various brain stem transection and lesion experiments (Jouvet 1962) and proposed that the pontine tegmentum was the source of signals that activated the cortex, moved the eyes, and shut off other portals of motor output in the brain stem and spinal cord (see again Fig. 3.4).

This sophisticated activation model is still valid and useful. Jouvet later discovered that cortical activation of sleep was not only tonic, but also phasic, and that the anterior pons contained the chemically coded neurons that could help us understand the chemical basis of differences between waking and dreaming consciousness.

It was then natural for Robert McCarley and me to seek the cellular basis of brain activation in sleep. Our initial studies, which used microelectrodes to record from individual brain cells, allowed us to identify and characterize numerous REM-on cell populations throughout the brain of cats. At first this work only suggested how the tonic REM sleep brain activation (A) process might be achieved. But later, our work (Nelson et al. 1983) led to the identification and cellular analysis of Jouvet's PGO wave system, an important contributor to the alterations in Factor I (information source) and to the chemical mechanisms by which the system changed state such that waking and dreaming consciousness were so psychologically different from each other (Factor M).

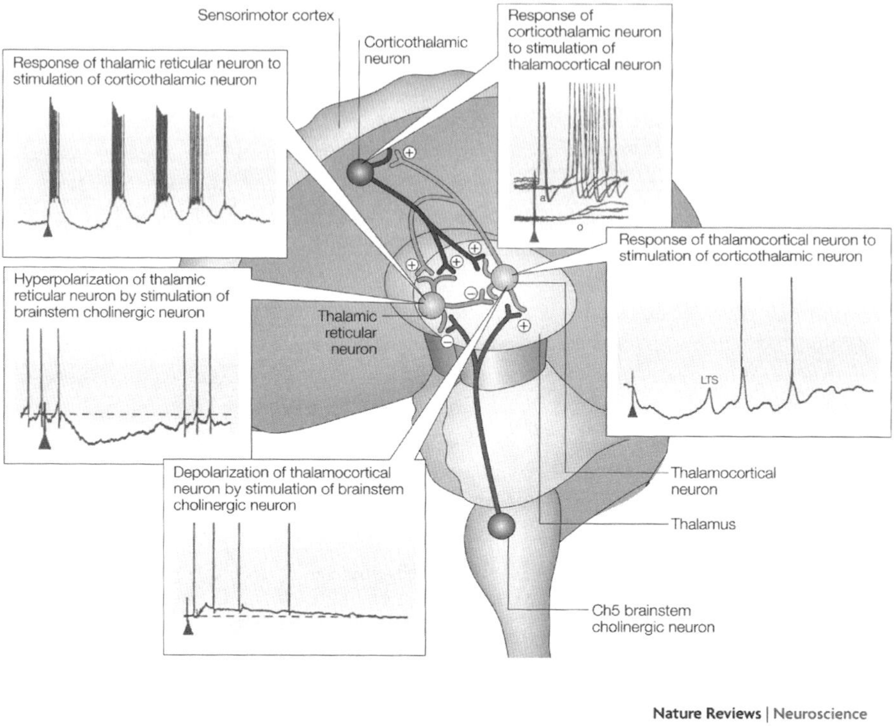

Fig. 3.3 The thalamocortical machinery for the generation of oscillatory rhythms of NREM sleep and associated plasticity processes. Structures involved in the production of thalamocortically generated non-rapid eye movement (*NREM*) sleep rhythms. These processes must be reversed by the activation mechanisms of waking and REM to facilitate consciousness. Anatomical structures with representative, schematically depicted neurons include the cholinergic pedunculopontine tegmental nucleus of the mesopontine brainstem, the reticular nucleus of the thalamus (which envelops the other thalamic nuclei), the combined specific (thalamocortical relay) and nonspecific (diffusely projecting) thalamic nuclei (mostly glutaminergic and excitatory), and the cortex (specific sensory regions of which are the targets of specific thalamic relay nuclei, whereas specific cortical motor regions project to different thalamic relay nuclei). Reticular thalamic neurons send inhibitory GABA (y-aminobutyric acid)-releasing projections to other thalamic neurons, whereas most thalamocortical and corticothalamic neurons send excitatory glutaminergic projections. Local thalamic and cortical inhibitory interneurons are not shown. Intracellular recordings depicted include: depolarizing (excitatory) effects of ascending cholinergic stimulation on excitatory thalamocortical neurons, in contrast to the inhibitory effect of such stimulation on inhibitory reticular thalamic neurons; excitatory effects resulting from stimulation of thalamocortical and corticothalamic neurons on each other (*a* antidromic action potential, *LTS* low-threshold spike, *o* orthodromic action potential in reponse to the same stimulus as a); and the characteristic spindle-frequency reponse of thalamic reticular neurons to excitatory corticothalamic stimulation (Reproduced, with permission, from Hobson and Pace-Schott 2002)

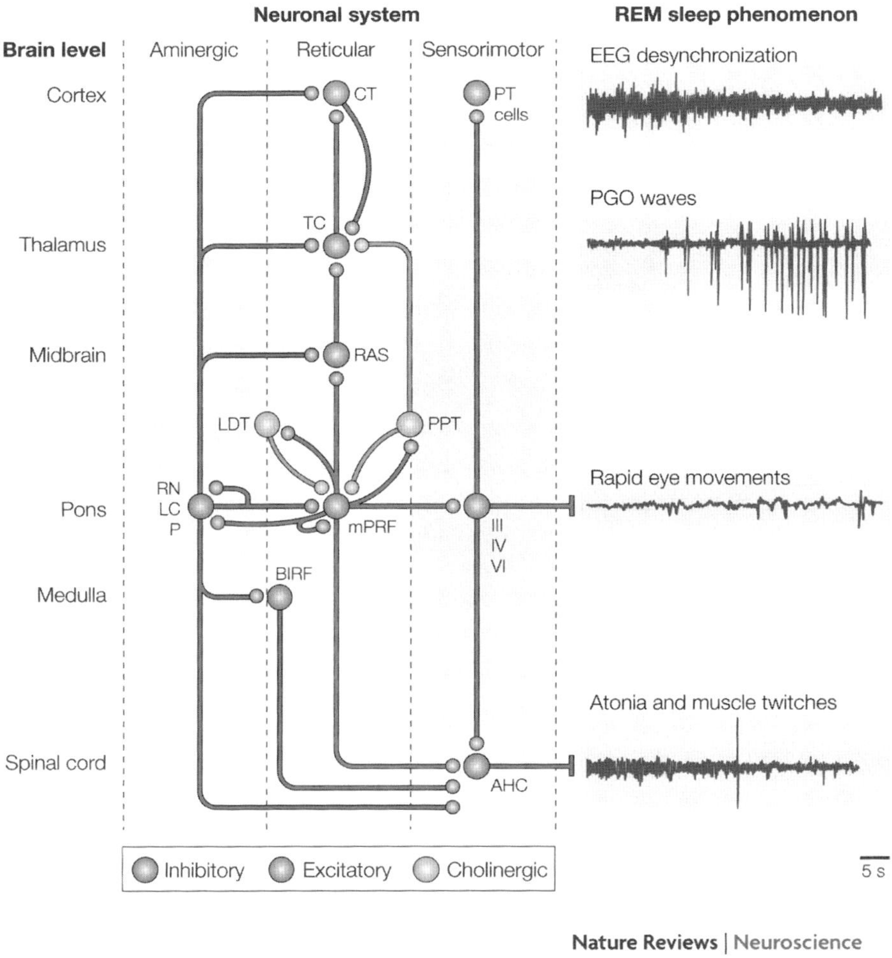

Fig. 3.4 Schematic representation of the process of REM sleep generation. The network comprises three neuronal systems (aminergic, reticular, and sensorimotor) that mediate rapid eye movement (*REM*) sleep electroencephalographic (*EEG*) phenomena (*right*). The actual synaptic signs of many of the aminergic and reticular pathways remain to be shown and, in many cases (such as the thalamus and cortex), the neuronal architecture is far more complex than is indicated here. (Compare with Fig. 3.3 to appreciate brainstem thalamocortical interaction in the activation of REM). *III* Oculomotor, *IV* trochlear, *VI* abducens, *AHC* anterior horn cell, *BIRF* bulbospinal inhibitory reticular formation (for example, gigantocellular tegmental field, parvocellular tegmental field, magnocellular tegmental field), *CT* cortical, *LC* locus coeruleus, *LDT* laterodorsal tegmental nucleus, *mPRF* meso- and mediopontine tegmentum (for example, gigantocellular tegmental field, parvocellular tegmental field), *P* peribrachial region, *PGO* ponto-geniculo-occipital, *PPT* pedunculopontine tegmental nucleus, *PT cell* pyramidal cell, *RAS* midbrain reticular activating system, *RN* raphé nuclei, *TC* thalamocortical (Reproduced, with permission, from Pace-Schott and Hobson 2002)

3.2 Information Source (Factor I)

The final demise of the deafferentation hypothesis was effected by the discovery that the brain actively *excluded* sensory stimulation in sleep by a combination of physiological mechanisms including disfacilitation, occlusion and postsynaptic inhibition.

At sleep onset, the reticulo-thalamic system becomes less capable of conducting sensory messages to the upper brain as it develops the oscillatory pattern of discharge seen in the EEG as spindles (in stages II and III) and as slow waves (in stages III and IV) of NREM sleep. The result is that more weakly excitatory inputs are effectively blocked from access to the brain's substrate of consciousness. On falling asleep we thus become isolated from the outside world by an active process. This is deafferentation but it is central, not peripheral and it is active, not passive (See, again, Fig. 3.4).

When intrinsic activation (A) begins to shift the brain from the depths of NREM sleep up to the heights of REM, three additional processes conspire to guarantee the brain's continued insensitivity to external stimulation. Presynaptic and postsynaptic inhibition of spinal reflexes and PGO wave generation will be considered as part of our discussion of Factor I, while aminergic demodulation will be considered later as part of our discussion of modulation (Factor M).

The late Ottavio Pompeiano of the University of Pisa in Italy pioneered the work leading me to conceive of Factor I. Pompeiano and Morrison discovered that each burst of eye movement during REM was associated with presynaptic inhibition of the group Ia afferent fibers from muscle spindles in the spinal cord connected to peripheral sensors in the periphery of the body (Morrison and Pompeiano 1965). Pompeiano also found that pre-synaptic inhibition quelled transmission in the caudate nucleus, the pathway to the upper brain (Morrison and Pompeiano 1966; Pompeiano 1967). The result is that the likelihood of an external sensory message getting into the brain is decreased just as the brain is being bombarded by intense stimulation by the internally generated PGO activity.

Using macroelectrodes, Jouvet had shown that the eye movements of REM sleep correlated with 250 microvolt amplitude EEG spike and wave complexes. These so-called PGO waves (visible in Fig. 3.4) denoted powerful internal excitation of the pontine tegmentum (P), the lateral geniculate body of the thalamus (G), and the occipital cortex (O) of the cat. We now know that comparable signals are widespread in posterolateral cortical regions of the human brain. These PGO waves are the best candidates, so far, for the long sought dream stimuli. PGO waves are internally generated messages arising in one part of the brain and conducted to other brain parts. Whatever their contribution to dreaming, PGO waves are without doubt potent internal sensory signals which convey a large amount of information about eye movement to the upper brain. As such, they must somehow contribute to our thinking about consciousness.

For my protoconscious theory, this is a particularly important fact. The REM sleeping brain is not only turned on (Factor A) but it sends itself messages that are essential to the sensorimotor integration underlying waking consciousness. We move our eyes continuously when we are awake and yet the world does not appear to move. A great deal of neuronal computation is required to explain this fact. And that computation goes on in sleep, as Shakespeare's sonnet 43 suggests, "When most I wink, then do mine eyes see best."

Shakespeare's Sonnet 43

When most I wink, then do mine eyes best see,
For all the day they view things unrespected;
But when I sleep, in dreams they look on thee,
And darkly bright are bright in dark directed.
Then thou, whose shadow shadows doth make bright,
How would thy shadow's form form happy show
To the clear day with thy much clearer light,
When to unseeing eyes thy shade shines so!
How would, I say, mine eyes be blessed made
By looking on thee in the living day,
When in dead night thy fair imperfect shade
Through heavy sleep on sightless eyes doth stay!
All days are nights to see till I see thee,
And nights bright days when dreams do show thee me.

This poem is designed to prove the author's love of his sweetheart but it also serves to emphasize the intensity of dream vision (Shakespeare 2008).

Single cell recording studies indicate that PGO waves arise in the far lateral anterior pons in the region of the brachium conjunctivum. Brain stem cells there fire in bursts. They are cholinoceptive and they can be activated by exogenous cholinergic agonists. They fire if, and only if, there is an ipsiversive eye movement 25 ms later. They thus convey to the thalamocortical system information about the motoric intent of the oculomotor system. According to my colleague, Subimal Datta, the neurons which convey PGO information to other parts of the brain are glutaminergic (Datta 1995).

PGO waves arise in the thalamic geniculate body 12.5 ms after pontine burst cell discharge begins. This is still 125 ms before the eye has begun to move! The waves in the ipsilateral geniculate are about twice the amplitude of those in the contralateral geniculate indicating a preferential ipsilateral target for the message of oculomotor intention.

We can thus say, for certain, that there is intense internal information generation regarding pending eye movements proceeding from the oculomotor region of the brain stem to sensory regions of the thalamus and cortex in REM sleep. These waves of information are similar to the spindles and slow waves of NREM sleep in that they constitute entirely automatic oscillations of neuronal activity. They occlude sensory input by occupying pathways which must be open to allow external information entry into the system. In an analogy to the telephone, the line is busy in REM sleep as far as receiving calls is concerned. To process external information, sensory signals from the external world must wait until the thalamocortical gating system is again open.

Factor I not only measures the degree to which the system is offline and occluded in the sensory sense. It also focuses on the degree to which internally generated motor signals–generated in response to the PGO like neuronal activity–are actively prevented from translation into actual movement. Understanding this active paralysis began in the work of Horace Magoun and Ruth Rhines (1947) which demonstrated that muscle tone could be abolished, and movement prevented, by electrical stimulation of the medullary reticular formation. The medullary reticular formation is just caudal to the cortical activation system of the midbrain and the REM generator system of the pons (See again Fig. 3.4).

Michel Jouvet and Francois Michel (1959) added active abolition of muscle tone to the epochs of REM with cortical activation that Dement and Kleitman (1957) had already described in cat sleep. By 1959, it was thus already clear that the brain could be turned on (activated, Factor A) and sleep preserved if motor output was blocked (input-output gates closed, Factor I). Ottavio Pompeiano and Adrian Morrison then showed that the inhibition of the anterior horn cells of the spinal cord was postsynaptic, a finding that was confirmed by Michael Chase and Francisco Morales at UCLA using intracellular probes that were capable registering the 10 mV of membrane hyperpolarisation that made motorneurons much more difficult to excite in REM sleep (Chase and Morales 1983).

The idea that the motor inhibition, like the REMs and the cortical activation waves, were of brain stem origin, was rendered more plausible when, first, Jouvet and Delorme, then Morrison and Henley, showed that that lesions of the pontine tegmentum in the vicinity of the locus coeruleus were often followed by REM sleep without atonia (Jouvet and Delorme 1965; Henley and Morrison 1974). The cats, which were in REM with PGO waves, then jumped to their feet and executed a complex motor sequence of attack and defense postures. An inescapable inference was that during REM sleep dreams such movements are commanded but they are normally prevented from expression by the motor inhibition component of Factor I. The implications for the motoric animation of dreaming are self-evident.

Protoconsciousness theory welcomes the functional significance of what would appear to be elaborate behavioral programming of instinctual acts such as escape, defense, and attack. The readout of these behaviors during REM sleep-without-atonia would seem to suggest that the brain is automatically programmed to emit important survival movements. These movements may normally be inhibited but can be experienced by us in dream consciousness when we flee imaginary attackers or turn and confront our imaginary aggressors (just as we may have to do in waking life). This aspect of the theory fits with other predictive or anticipatory features of our virtual and actual behavioral repertoires. This component of the protoconsciousness hypothesis echoes the earlier notions of REM-dream function voiced by Michel Jouvet and more recently enunciated as threat avoidance theory by Antti Revonsuo (Revonsuo 2000).

3.3 Modulation (Factor M)

The location of the pontine cell bodies of neurons which manufacture the neuromodulators norepinephrine (NE) and serotonin (5-HT) was established by the Swedish neuroanatomists working with Kjell Fuxe in the early 1960s (Dahlstrom and Fuxe 1964). The diminished output of these chemically coded cells in NREM sleep and their virtual extinction during REM were described in 1975 when McCarley and I defined them as "REM-off" cells (Hobson et al. 1975). The important implication of these findings for sleep and dream theory was that the activation (A) and offline status (I) of the brain was associated with, and perhaps caused by, aminergic demodulation (M).

At the same time that aminergic activity declined in REM sleep, cholinergic activity increased. This idea gave rise to the reciprocal interaction model of sleep cycle control and set the stage for the activation-synthesis theory of dreaming (Hobson and McCarley 1977). Testing these two theories has revealed that both REM sleep and dreaming are enhanced by cholinergic agonists. In humans, pilocarpine injected intravenously during the first NREM period induces earlier and longer first REM periods. These cholinergically enhanced episodes are associated with dreaming. Extensive study of cholinergic REM enhancement in cats by Helen Baghdoyan revealed a REM sleep trigger zone in the paramedian pons (Baghdoyan et al. 1989) while José Calvo and Subimal Datta defined a REM sleep regulatory region in the lateral pons. The two regions are connected mono-synaptically as demonstrated by James Quattrochi.

Indirect evidence for an active role of norepinephrine, serotonin, and histamine in waking is extensive and goes a long way toward helping us to explain the formal differences between dreaming and waking consciousness discussed in Lecture I. Most prominent is the dependence upon serotonin for recent memory, a key executive ego function. Also significant is the dependence of attention on norepinephrine. The dreamer, who cannot remember in his dreams, also cannot perform such important cognitive functions as orientation, logical thinking, and self-reflective awareness. As a consequence, insight and judgment suffer in dreaming. Unable to pay attention, the dreamer does not even notice his glaring cognitive defects. The dreamer has a biochemically determined agnosia which is miraculously ended by waking up! Waking is associated with the turning back on of the aminergic neuromodulatory neurons and a consequent renewed perfusion of the brain with serotonin, norepinephrine and histamine.

Acetylcholine (ACh) and dopamine (DA) release are not suppressed in REM sleep. If anything, they may be more potent than they are in waking because while ACh release (measured by microiontophoresis) is at about same level in waking as it is in REM sleep, it is more pulsatile in its delivery pattern and, like glutamate, it is more phasically exciting to its post-synaptic neuronal domain. That fact suggests a stimulus-like role for acetylcholine in dreaming consciousness. Thus it could be responsible for the burst-like firing, in REM, of lateral geniculate body and cortical neurons. In the new AIM model, these internal stimuli are hypothesized to be the pseudosensory signals that make dreaming so intensely visual. The Shakespearean sonnet question is answered and the answer, sadly, has nothing to do with love. The eye sees best when I wink most because it is then that I am in REM sleep.

Dopamine and acetylcholine are both major players in normal motor control. It is thus plausible that the offline running of motor programs, emphasized by the activation-synthesis theory of dreaming, is supported by the abundant availability of these two neuromodulators in REM sleep. The pulsatile release of ACh and DA in the thalamus and cortex, and other subcortical motor areas, could be related to the subjective experience of motoric animation in REM sleep dreams. The fact that experimental animals with pontine lesions, and patients with REM sleep behavior disorder (RBD), act out their dreams is significant because both predisposing neurobiological conditions involve alteration in the balance of the aminergic and cholinergic systems.

The connection between motor pattern generator neurons and neuromodulatory elements is disrupted by pontine lesions and the synaptic efficacy of motor inhibition is disrupted in the case of REM sleep behavior disorder. The net effect, in both cases, is a tendency for motor commands, normally suppressed in REM sleep by active motor inhibition, to be acted out, often awakening the animal or human subject.

Like dream perception, dream emotion may be affected by the change in aminergic-cholinergic balance. Just as there is cholinergic potentiation in sensorimotor regions of the forebrain, there is selective activation of the limbic regions of the emotional brain. The deep frontal white matter, the amygdala, the anterior cingulate, and the parahippocampal cortex are all selectively activated in REM. PGO waves can be recorded from the amygdala in REM sleep and amygdala stimulation can evoke PGO waves in the geniculate during the NREM to REM transition. It is therefore likely that the subjective experience of the three leading dream emotions, anxiety, elation and anger, may be related to phasic activation of limbic lobe structures in REM sleep.

Protoconsciousness theory takes account of the selective activation of these important survival emotions in REM sleep. The brain-mind is prepared to respond with fright, flight, and fight (or sometimes with approach) when stimuli evoke one or the other set of responses in waking. We are driven away or drawn closer to others by environmental signals and we already know instinctively how to respond to those signals. Socially useful emotions such as shame, guilt, and sadness may be learned rather than pre-programmed because they are not essential to survival.

I have already pointed out that dream cognition, however bizarre, is usually congruent with dream emotion. This salience is robust enough to lead us to consider that dream emotions may shape, or even trigger, dream cognition. Dreaming may be ‘crazy’ in its hallucinations and delusions, its incongruity and its discontinuity, but it does make sense emotionally. This could be good news for the dream interpreters, but a word of caution is indicated.

The emotion dream-cognition link may be neither as unique nor as profound as most psychological dream theories suppose. Furthermore dreams may not be *uniquely* informative about an individual’s associations to emotional stimuli. The range of dream emotions is so surprisingly limited as to make us wonder whether dreaming consciousness is not a relatively unfavorable ground for psychodynamic investigation. In other words, waking experience may tell us all we need to know about emotional salience. For all those therapists who have traveled Freud's Royal Road to the Unconscious, this suggestion will be rejected as heretical.

The unfettered play of dopamine in REM sleep is in keeping with the assumption that dreaming is "motivated" and that important motivational goals may be revealed in dreams. Dreams are not entirely meaningless and there is still reason for considering them as uniquely hypermeaningful. This is an unproved hypothesis which must be questioned, however, until studied experimentally. Why dreaming should reveal a deeper, more profound, truth about our intentions is not yet clear and while we may want to believe such a thing, we must remain skeptical.

As a psychotherapist with over 40 years of experience I can honestly assert that I have never learned anything from a patient's dream that I didn't already know. This conclusion is shared by anti-Freudian dream research experts, like William Domhoff, who claim that dreamers have the same set of psychological concerns and styles in dreaming as they do in waking. While I share his anti-Freudianism, I am skeptical of Domhoff's continuity hypothesis (Domhoff 2002).

Nightmare.

The attack of the fascist airplanes on the village of Guernica inspired Pablo Picasso to paint a picture that has been recreated by Jay Vogelsong to depict dream aggression and anxiety.

3.4 Summary of Physiological Underpinnings of Dream Consciousness

I conclude that the main similarities and differences between dreaming and waking consciousness mirror changes in the physiological condition of the brain as follows:

1. There is indeed a brain activation process associated with both states. Hence, some form of consciousness is possible in both waking and dreaming.
2. The brain-mind is offline in REM, because external sensory inputs and motor outputs are blocked. The activated brain processes stimuli of its own devising. Being primarily visuomotor, these internal stimuli render dream consciousness more vividly perceptual than waking but because of the exclusion of space-time data from the outside world, the percepts are often characterized by incongruity and discontinuity.
3. Cognitive processes such as memory, self-reflective awareness, insight and judgment are deficient in dream consciousness owing to the shift in balance between the aminergic system (which is dominant in waking but ineffective in REM) and the cholinergic system (which is suppressed in waking but unfettered in REM).

3.5 The AIM Model

Having classified the psychophysiology of the brain-mind system according to the three major functions, activation (A), input-output (I), and modulation (M), it is natural to consider the dynamics of interaction between these three variables. A, I, and M can each be quantified and each set of the three values can be represented as a single point. One interesting way of doing this is to create an artificial, three dimensional state space, with A, I, and M values laid out as three orthogonal dimensions as in Fig. 3.5.

The right half of the state space, with at least 50 % values of A max is the locus of normal conscious state vicissitudes. At values of A that are less than 50 % of A max consciousness is minimal or nonexistent because the system does not have enough energy to sustain the neuronal drive and the functional coordinates necessary to support waking or dreaming consciousness. Note that activation can be high with contrasting values of I and M. If one displays arbitrary values for A, I, and M, then loci emerge as domains corresponding to the cardinal states of waking, sleeping and dreaming as shown.

The waking domain is in the right upper rear corner of the state space because A is high (right third), I is high (rear third) and M is high (upper third). There are numerous substates of waking which invite consideration using this paradigm. They range (in intensity of A) from hyperalert arousal to extreme drowsiness. They range (in intensity of I) from exclusively exteroceptive as in vigilant scanning of the horizon to exclusively interoceptive as in daydreaming. If M is also high then

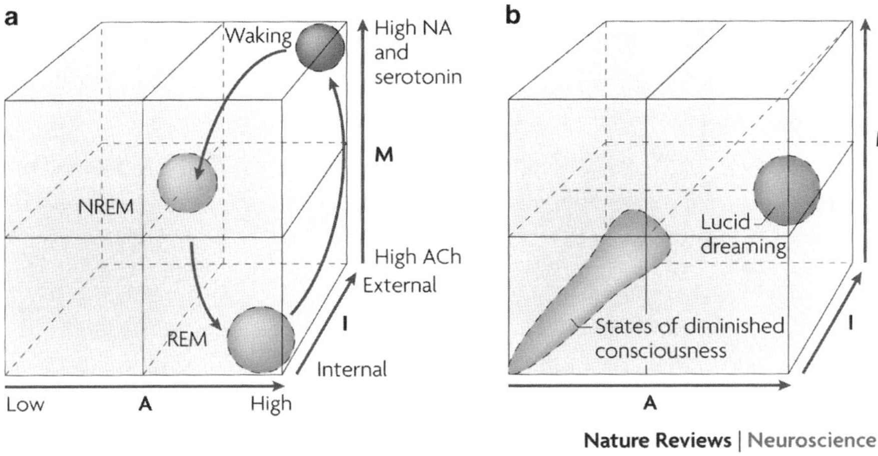

Fig. 3.5 AIM model of brain–mind state control. (**a**) The three-dimensional AIM state-space model showing normal transitions within the AIM state space from waking to non-rapid eye movement (*NREM*) and then to rapid eye movement (*REM*) sleep. The x axis represents A (for activation), the y axis represents M (for modulation) and the z axis represents I (for input–output gating). The values of A, I, and M can be derived from the neuronal data of animal experiments; factors A and I can also be estimated in human sleep laboratory data but, as yet, there is no way of measuring factor M in humans. Waking, NREM sleep, and REM sleep occupy distinct loci of this space. Waking and REM sleep are both in the right-hand segment of the space, owing to their high activation levels, but they have different I and M values. Thus, the activated, REM-sleeping brain–mind is both off-line and chemically differentiated compared with the waking brain–mind. NREM sleep is positioned in the centre of the space because it is intermediate in all quantitative respects between waking and REM sleep. The values of A, I, and M change constantly, but the changes are constrained. Sleep and waking states alternate owing to circadian influences (not shown). During sleep, AIM values tend to follow elliptical trajectories through the space. As sleep advances in time, AIM values go less deeply into the NREM sleep domain and more deeply into the REM sleep domain. The normal, cardinal domains of waking, NREM, and REM sleep occupy relatively limited zones of the space. (**b**) Diseases, such as those neurological conditions that produce coma and minimally conscious states, can be arrayed in the left-hand segment of the space, owing to their low activation values. Lucid dreaming, which is a hybrid state with features of both waking and dreaming, may be situated in the middle of the extreme right-hand side of the AIM state space between waking and REM, towards either of which lucid dreamers are drawn. Sleep and psychiatric disorders can ultimately be placed in such a schema. *ACh* acetylcholine, *NA* noradrenaline, *5-HT* serotonin (Reproduced, with permission, from Hobson 2009)

exteroception is favored via the enhancement of attention. This might occur when vigilance is the appropriate response to a possible threat. If attention abates and comes to be directed by another person, then dissociated states, like hypnosis, may be favored. Further discussion of waking substates is of obvious importance in reconceptualizing normal mental function in physiological terms.

When A declines and input-output gates (I) begin to close, M also declines and the AIM system enters the NREM sleep domain in the center of the state space. In the depths of stage IV NREM sleep about 50 % of A max, activation is at its lowest normal ebb. Consciousness, if it exists at all, is minimal in NREM sleep and it is restricted to perseverative thoughts. It is very difficult to escape from this trough of brain deactivation. Attempts to arouse experimental subjects from NREM sleep and my own personal experience with early night phone calls yield awakenings which are confusional as well as difficult to sustain. There is often memory and information loss as if the reduced levels of M were low enough to render the mind practically useless for the performance of wake-state problem-solving tasks.

As NREM sleep gives way to REM sleep, the brain reactivates. Factor A goes back up to wake state levels or in some regions of the brain even higher. But now the input-output gates close completely and Factor I goes to its lowest levels (with respect to external access). Because A is high, but both I and M are low, the REM sleep domain is in the right, lower and anterior corner of the state space.

Together with this closure of sensory and motor gates, the offline brain-mind system begins to elaborate and process its own intrinsic signals, such as the information carried by the PGO waves. As described above, these signals arise from the oculomotor system of the brain stem and inform both the thalamus and cortex about the direction and amplitude of upcoming eye movements. Whether and how these gaze codes are used to construct dream imagery is unknown but, from the time of Helmholtz, the existence of feed-forward efferent copy data have been thought to be important for sensorimotor integration in waking. Since sensorimotor integration is also clearly involved in the synthesis of dreams, it stands to reason that such precise internally generated data would be used in dream image production. Helmholtz himself suggested as much in his classic work, *Physiological Optics* (von Helmholtz 2000).

The fourth dimension of the AIM model is time and in a night of human sleep there may be four (or five) elliptical trajectories of AIM beginning in the waking domain, proceeding slowly through the NREM sleep domain to REM sleep and then swinging rapidly back up toward waking for another cycle. The result is a snail-like pattern of ellipses in the state space as can be seen in other representations of AIM.

Each cycle is about 90–100 min in duration in the adult human. There is progressively less time spent in the NREM domain and correspondingly more time spent in the REM domain in each successive cycle. This progression is in keeping with the known propensity for deep NREM sleep to dominate the first half of each night and for REM sleep to dominate the second half of the night. Despite the fact that, overnight, sleep is getting steadily lighter with respect to itself in each successive cycle, REM sleep continues to be more highly correlated with hallucinogenic dreaming than does NREM sleep.

The upshot of this new way of looking at brain-mind state is to recognize its integrated nature, its complexity and its simplicity, and above all its continuous dynamism. There are probably many more dimensions to the state space model than the four that I have emphasized but four are already twice as many as two. The opportunity to map dream consciousness psychology and neurophysiology at the cellular and molecular levels are strong aspects of the model.

Another important aspect of the model concerns the rest of the state space defined by the model. Problematical and pathological domains abound as do reversals of the normal temporal dynamics of the trajectories. For example, the low activation domain of the left half of the space may be occupied by the brain-mind following stroke or lesion damage to the brain stem. In narcolepsy, subjects may enter the REM domain directly from waking (as is the case of younger human subjects who, like Eugene Aserinsky's son, Armand, experience sleep onset REM periods). Unusual conditions like lucid dreaming, which contains aspects of both waking and dreaming, can also be located between the REM and wake domains. With AIM, we have a theoretical framework for all brain-mind states both pathological and normal.

3.6 Human Neuropsychology

The AIM model, while valid for the brain as a whole, does not take into account the regional differences in activation or deactivation and therefore may be seen to be based upon the presumption that all regions are equally activated and deactivated over the sleep cycle. (Ideally, an AIM model might some day be constructed for each brain region). But we already know from PET and MRI imaging studies, as well as from microelectrode experiments, that this isn't the case (Braun et al. 1997; Maquet et al. 1996). Compared to waking, regional brain excitation can be globally decreased in (NREM) sleep and selectively reactivated (or even hyperactivated) in REM sleep or not reactivated at all in REM (Fig. 3.6).

The resulting differentiated pattern is of enormous consequence for our effort to contrast dreaming and waking consciousness. Regions which are hyperactivated in REM sleep (compared to waking) include the pons (whose selective activation and internal role in the brain activation of sleep has until now depended upon microelectrode studies in animals), the parietal operculum (known to be involved in spatial integration), the amygdala (known to be a limbic lobe mediator of emotion, especially anxiety) and the deep frontal white matter (known to be involved in the integration of emotion and cognition). Two of these four areas, the parietal operculum and the deep frontal white matter, appear to be essential to dreaming on the basis of claims of cessation of dream consciousness by patients with strokes localized to those regions.

One cortical brain area, the dorsolateral prefrontal cortex (DLPFC) is thought to be involved in executive ego functions. It is not reactivated in REM sleep, meaning that it is deactivated compared to waking. The specific and robust cognitive defects that have been found to characterize dream consciousness may be lost as a consequence of this failure to reactivate the DLPFC in REM. The loss of precisely such cognitive functions as working memory, self-reflective awareness, volition and planning supported by the DLPFC may be explained by the failure to reactivate it in REM sleep.

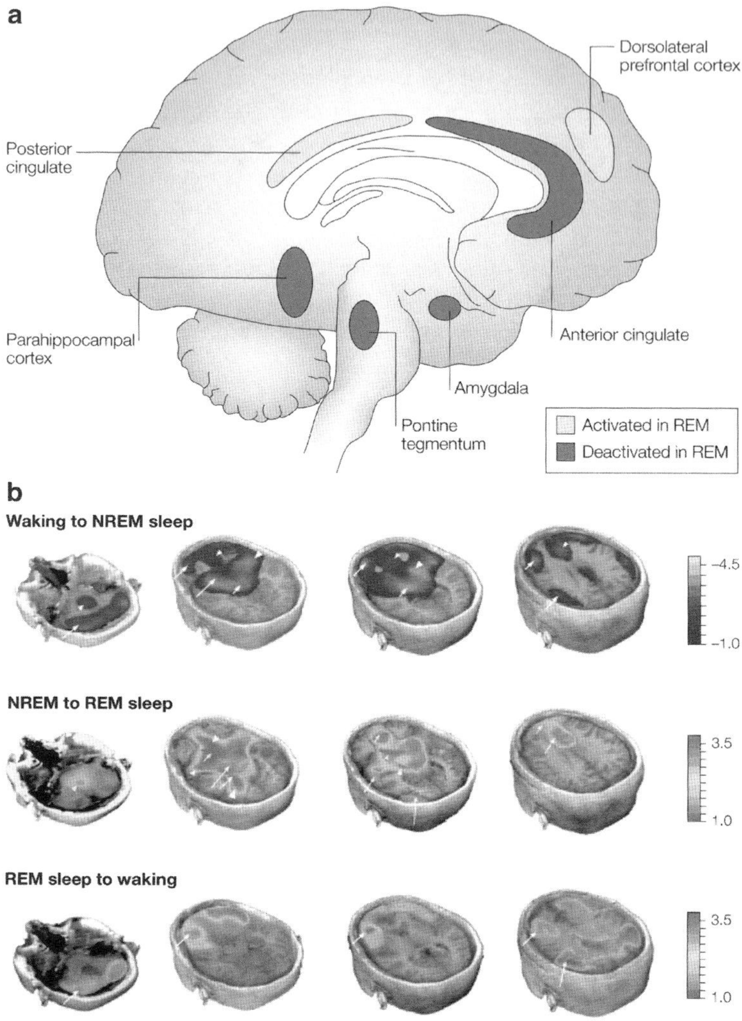

Fig. 3.6 Brain activation during sleep and waking. (**a**) Sagittal view of the human brain showing areas that were activated or deactivated in rapid eye movement sleep (*REM*) compared with waking and/or non-REM (*NREM*) sleep in two or more of three positron emission tomography (*PET*) studies. A schematic (rather than a morphologically realistic) view is shown of only the areas that could be easily matched between two or more studies. Considerably more extensive areas of activation and deactivation are reported in individual studies. The depicted areas are, therefore, representative portions of larger areas that subserve similar functions (such as limbic-related cortex, ascending activation pathways and multimodal association cortex). (**b**) Successive horizontal sections of the brain showing changes in relative activity between waking and NREM, NREM and REM, as well as REM and waking, using H_2O PET. Z-score contrast between the respective pairs of behavioral states are indicated. Values are z-scores that represent the significance level of changes in regional cerebral blood flow at each voxel (Figure reproduced from Braun et al. 1997) (Color figure online)

These correlations are almost too good to be true, and a detailed causal analysis waits upon a second generation of experimental work. This will be difficult to do even if the experiments are considered worthwhile by granting agencies, which are rapidly shifting their favor toward molecular biology. It is one purpose of this lecture to argue strongly for such work however costly and risky it may be. The differentiation of dreaming and waking consciousness is an experiment of nature which must be of profound functional significance as well as offering a unique opportunity to use the dream-wake paradigm as a way of gaining further insight about the nature and mechanism of brain-mind integration.

There is no other example of such strong correlation between brain and mind as that which has been demonstrated over the wake-sleep-dream cycle. Given the fact that all that we know has been accrued in only 60 years, we can hope to learn much much more during the next century. An important point to stress is that not all the progress that we can expect necessitates high tech studies such as microelectrode recordings for individual nerve cells, microinjection pharmacology and PET-MRI scanning.

There is, of course, a great deal of work remaining to be done in these domains. But for many important and detailed studies of the formal aspects of mentation, much less demanding technology is sufficient. Complementing the human sleep lab are home based recording techniques such as portable EEG and non-analogous devices such as the Nightcap (see again Fig. 2.3).

What is most important to grasp is that the paradigm shift from content interpretation to form quantification puts dream and related consciousness research within reach of scientifically motivated psychology departments in colleges (or even high schools) if only prospective investigators are rigorous and well-informed about neurobiology and consider the formal approach to subjective experience.

We can now say that what we know of the neurophysiology of REM sleep is entirely consistent with the protoconsciousness hypothesis. Brain activation of the fetus is instantiated as early as the third trimester of pregnancy (See Figs. 4.1a, 4.1b, and 4.2). This activation is surprisingly stereotyped. REM sleep constitutes an elaborate program for sensorimotor integration, emotion evocation, and scenario construction. It may be hard to prove but it is hard to believe that these features are not in the service of waking consciousness. The fact that this state recurs every night of our lives underscores that assertion.

References

Aserinsky, E., & Kleitman, N. (1953). Regularly occurring periods of ocular motility and concomitant phenomena during sleep. *Science, 118*, 361–375.

Baghdoyan, H. A., Lydic, R., Callaway, C. W., & Hobson, J. A. (1989). The carbachol-induced enhancement of desynchronized sleep signs is dose dependent and antagonized by centrally administered atropine. *Neuropsychopharmacology, 2*(1), 67–79.

Braun, A. R., et al. (1997). Regional cerebral blood flow throughout the sleep–wake cycle. *Brain, 120*, 1173–1197.
Bremer, F. (1935). Cerveau isolé et physiologie du sommeil. *Comptes Rendus de la Societe de Biologie (Paris), 118*, 1235–1241.
Chase, M. H., & Morales, F. R. (1983). Subthreshold excitatory activity and motorneuron discharge during REM periods of active sleep. *Science, 221*(4616), 1195–1198.
Dahlstrom, A., & Fuxe, K. (1964). Evidence for the existence of monoamine-containing neurons in the central nervous system. I. Demonstration in the cell bodies of brain stem neurons. *Acta Physiologica Scandinavica, 62*, 1–55.
Datta, S. (1995). Neuronal activity in the peribrachial area: Relationship to behavioral state control. *Neuroscience and Biobehavioral Reviews, 19*(1), 67–84.
Dement, W. (1958). The occurrence of low voltage, fast, electroencephalogram patterns during behavioral sleep in the cat. *Electroencephalography and Clinical Neurophysiology, 10*(2), 291–296.
Dement, W., & Kleitman, N. (1957). Cyclic variations in EEG during sleep and their relation to eye movements, body motility, and dreaming. *Electroencephalography and Clinical Neurophysiology, 9*(4), 673–690.
Domhoff, G. W. (2002). *The scientific study of dreams: Neural networks, cognitive development and content analysis*. Washington, DC: American Psychological Association Press.
Henley, K., & Morrison, A. R. (1974). A re-evaluation of the effects of lesions of the pontine tegmentum and locus coeruleus on phenomena of paradoxical sleep in the cat. *Acta Neurobiologiae Experimentalis (Wars), 34*(2), 215–232.
Hobson, J. A. (2009). REM sleep and dreaming: Towards a theory of protoconsciousness. *Nature Reviews Neuroscience, 10*, 803–814.
Hobson, J. A., & McCarley, R. W. (1977). The brain as a dream state generator: An activation-synthesis hypothesis of the dream process. *American Journal of Psychology, 134*(12), 1335–1348.
Hobson, J. A., & Pace-Schott, E. F. (2002). The cognitive neuroscience of sleep: Neuronal systems, consciousness and learning. *Nature Reviews Neuroscience, 3*(9), 679–693.
Hobson, J. A., McCarley, R. W., & Wyzinski, P. W. (1975). Sleep cycle oscillation: Reciprocal discharge by two brainstem neuronal groups. *Science, 189*(4196), 55–58.
Jouvet, M. (1962). Research on the neural structures and responsible mechanisms in different phases of physiological sleep. *Archives Italiennes de Biologie, 100*, 125–206.
Jouvet, M., & Delorme, F. (1965). Locus coeruleus et sommeil paradoxal. *Comptes Rendus de la Societe de Biologie, 159*, 895–899.
Magoun, H. W., & Rhines, R. (1947). An inhibitory mechanism in the bulbar reticular formation. *Journal of Neurophysiology, 9*, 165–171.
Maquet, P., Péters, J., Aerts, J., Delfiore, G., Degueldre, C., Luxen, A., & Franck, G. (1996). Functional neuroanatomy of human rapid-eye-movement sleep and dreaming. *Nature, 383* (6596), 163–166.
Morrison, A. R., & Pompeiano, O. (1965). Central depolarization of group Ia afferent fibers during desynchronized sleep. *Archives Italiennes de Biologie, 103*, 517–537.
Morrison, A. R., & Pompeiano, O. (1966). Vestibular influences during sleep III. Dissociation of the tonic and phasic inhibition of spinal reflexes during desynchronized sleep following vestibular lesions. *Archives Italiennes de Biologie, 104*, 321–346.
Moruzzi, G., & Magoun, H. W. (1949). Brainstem reticular formation and activation of the EEG. *Electroencephalography and Clinical Neurophysiology, 1*, 455–473.
Nelson, J. P., McCarley, R. W., & Hobson, J. A. (1983). REM sleep burst neurons, PGO waves, and eye movement information. *Journal of Neurophysiology, 50*(4), 784–797.
Pace-Schott, E. F., & Hobson, J. A. (2002). The neurobiology of sleep: Genetics, cellular physiology and subcortical networks. *Nature Reviews Neuroscience, 3*, 591–605.

Pompeiano, O. (1967). The neurophysiological mechanisms of the postural and motor events during desynchronized sleep. *Research Publications – Association for Research in Nervous and Mental Disease, 45*, 351–423.

Revonsuo, A. (2000). The reinterpretation of dreams: An evolutionary hypothesis of the function of dreaming. *Behavioral and Brain Sciences, 23*(6), 877–901.

Shakespeare, W. (2008). In C. Burrow (Ed.), *The Oxford Shakespeare: The complete sonnets and poems*. New York: Oxford University Press.

Sherrington, C. S. (1906). *Integrative action of the nervous system*. London: Constable.

Steriade, M. (1981). Mechanisms underlying cortical activation: Neuronal organization and properties of the midbrain reticular core and intralaminar thalamic nuclei. In O. Pompeiano & M. C. Ajmone (Eds.), *Brain mechanisms and perceptual awareness*. New York: Raven.

von Helmholtz, H. L. F. (2000). Helmholtz's treatise on physiological optics (3 Vols). Translated from the 3rd German ed. In: J. P. C. Southall (Ed.) London: Thoemmes Press (Back In Print) 1962.

Watson, J. D., & Crick, F. H. (1953). Molecular structure of nucleic acids: A structure for deoxyribose nucleic acid. *Nature, 171*(4356), 737–738.

Chapter 4
Lecture III: Philosophy

J. Allan Hobson

From Matter into Mind

I found the universe within my brain—
Not writ in code or mirrored in a thought,
But real, as flesh: in atoms I retain
Of stardust and the start of time from naught.
Yes, underneath this integrative state
In which I know I "am" are but the chance
Selections and connections that collate
Inherent values with found circumstance.
For, all this matter generating "mind"
Evolved (from out a mindless past) as means
To map environment: to match and bind
In memory how each perceived thing seems.
I call this firmament of function "me"
(Though substance-wise, no separate entity).

–David Borodin

While still highly correlational, and grossly incomplete, the documentation of the subjective changes that we experience when we dream and when we are awake are already strongly consonant with known changes in brain activity. Of course we would like the goodness of fit between psychology and physiology to be even more complete, and no doubt, in time it will be. Nevertheless, it is not too early to take account of what this apparent brain-mind unity means to our philosophy (defined as our way of thinking about such matters as the unconscious, the self, and mental illness).

Before taking up some of these philosophical issues, it is important to point out that, in addition to the microscopic details of physiology discussed in Lecture II, there are two major biological considerations which bear heavily upon our considerations

J.A. Hobson (✉)
Department of Psychiatry, Harvard Medical School, Boston, MA, USA
e-mail: allan_hobson@hms.harvard.edu

N. Tranquillo (ed.), *Dream Consciousness*, Vienna Circle Institute Library 3,
DOI 10.1007/978-3-319-07296-8_4,

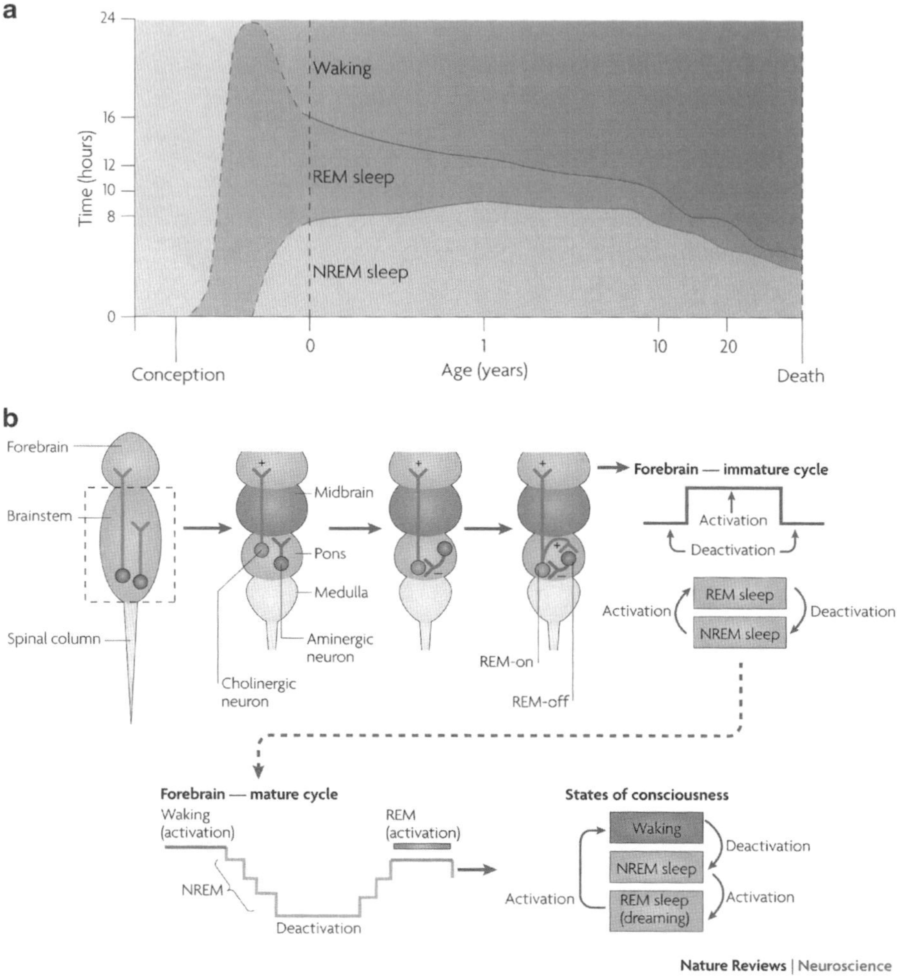

Fig. 4.1 Developmental considerations: Human sleep and age. (**a**) The marked preponderance of rapid eye movement (REM) sleep in the last trimester of pregnancy and the first year of life decreases progressively as waking time increases. Note that non-rapid eye movement (NREM) sleep time, like waking time, increases after birth. Despite its early decline, REM sleep continues to occupy approximately 1.5 h per day throughout life. This suggests that its strongest developmental contribution is to early brain–mind development but that it subsequently plays an equally indispensable part in brain–mind maintenance (Reproduced, with permission, from Hobson 2009). (**b**) Proposed steps in brain development leading to the emergence of the NREM–REM sleep cycle and its alternation with waking. Following fertilization, blastulation, gastrulation and neural crest formation (not shown) the developing brain takes on a recognizable form in a tripartite structure (forebrain, brainstem and spinal cord) and then through a second tripartite division of the brainstem (to form the midbrain, pons and medulla). Based on inferences derived from adult neurobiology, the figure schematizes how the establishment of neuronal location (in the brainstem) and chemical signature (aminergic or cholinergic) might parallel the development of, first, activation and inactivation and, later, of rapid eye movement (REM) sleep, non-rapid eye

of consciousness and how we conceptualize its relationship to brain functions. As alluded to in the Introduction, they are that while REM sleep evolved together with the thalamocortical brain relatively recently in evolutionary time, REM sleep occupies an inordinate amount of time in the early development of those species that evince it. These important features are illustrated in Figs. 4.1a and 4.2 which supply in their captions technical details for the interested reader to peruse.

Examination of Fig. 4.1a leads to the recognition of the strong dynamics of developmental changes in conscious state in humans. Before we are born, and to the extent that we are born prematurely, our first state is one of REM-like brain activation. This temporal priority of REM is prime evidence for the protoconsciousness hypothesis. It is simply a scientific fact that much early brain development occurs in sleep and that such development is organized and ordered in a specific manner. Since prenatal and early brain development occurs mainly in sleep, a functional hypothesis for developmental neurobiology is emergent. I suggest that such a development as protoconsiousness is the functional tie to waking.

With respect to the evolutionary picture, Fig. 4.2 makes clear that REM sleep evolved twice, once in birds and a second time in mammals. These two classes of animals are not genetically related but they share two related features: complex brains and thermoregulation. The protoconsciousness hypothesis recognizes the functional significance of these properties by postulating that more complex brains support greater cognitive and sensorimotor integration than simpler ones. More specifically, thermoregulation is a necessary adaptation which allows the more complex brain of birds and mammals to function properly. The role of REM sleep in promoting brain development is paralleled by a role of REM in the preservation of thermoregulatory capability.

4.1 Dual-Aspect Monism

The physiological findings that I have reviewed in Lecture II have impact upon formal philosophical systems, especially those entailed by the philosophy of mind. This area of academic inquiry is already very busy with the revisions forced by

Fig. 4.1 (continued) movement (NREM) sleep and waking. For example, initial activity of cholinergic brainstem neurons probably leads to forebrain activation; subsequent activity of aminergic brainstem neurons and their interaction with cholinergic neurons result in a primitive cycle of brain activation and deactivation. As a differentiated neuronal circuitry is formed in the forebrain, a complex set of features comes to characterize the REM–NREM sleep cycle. The protoconsciousness that I intuit would begin here and continue as REM sleep dreaming throughout life. During the third trimester of pregnancy in humans REM-like brain activation predominates, but in the first years of life the time that the brain is in the REM sleep state declines, whereas the time it is in the NREM sleep state and the waking state increases. These are the building blocks of adult conscious states, although many of the details of the transition to adult states of consciousness remain to be established (Reproduced, with permission, from Hobson 2009)

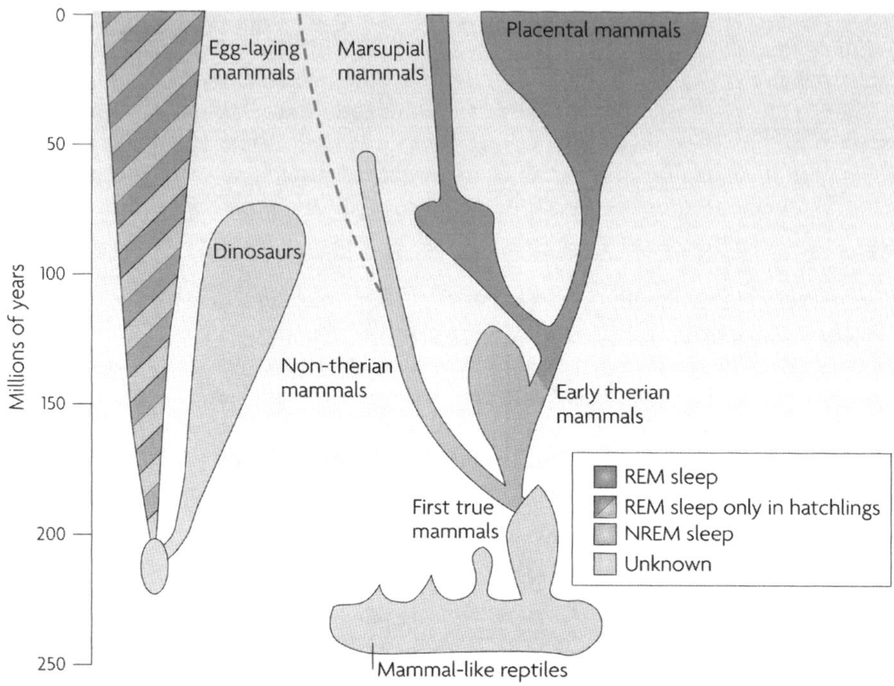

Fig. 4.2 Evolutionary considerations: The evolution of REM sleep. Birds and mammals evolved separately after branching off from the ancestral tree many millions of years ago. Both birds and mammals are homeothermic, and both classes of animal have appreciable cognitive competence. Restoration of thermoregulatory capacity is a strongly evidenced function of REM sleep. With respect to the possible enhancement of cognitive skills by REM, it is significant that both birds and mammals are capable of problem solving and both can generate forms of verbal communication (Reproduced, with permission, from Hobson 2009) (Color figure online)

modern neuroscience. I will argue that the data combined in the AIM model are most compatible with what philosophers have called dual-aspect monism. Dual-aspect monism holds that the brain as an object can be seen and measured by traditional third person scientific methods but that it is unique in possessing subjectivity, the famous "qualia," and self-awareness that make our waking conscious experience so unique. Qualia are first person experiences. They cannot yet be reduced to third person measurement and are hence outside the realm of science as we know it. The "hard problem" of philosophy cited by David Chalmers is to imagine how an object, the brain, can possess awareness. Some philosophers even believe that this problem is so hard that it will never be solved. I beg to disagree and have developed this conviction in two recent papers (Hobson and Friston 2012, 2014).

I will argue that the scientific study of waking, sleeping and dreaming has already gone a very long way to softening the hardness of the problem and can be expected, even with existing technology, to do much more. With as yet unforeseen technological developments, even more progress can be expected. In my most optimistic moments, I even assert that the mind-body problem is already solved! The rest of the work is a clean-up operation. For me, the brain is a subjective object and its states are objective subjects. This simple idea is discussed in more detail in my book, *Dream Life: An Experimental Memoir* (Hobson 2011).

Whatever your verdict may be about this bold claim, I hope to convince you of two further points:

- The *first point* is that a science of first person experience is both desirable and feasible. I will develop the thesis that a responsible approach to introspective phenomenology must and can be developed. Now that we have third person data acting as a scaffold upon which to pin subjectivity we are protected from the unchecked speculation of subjectivity that gave first person evidence such a bad name up to and through the nineteenth century. My own strategy for the scientific study of the mind in relation to the activity of the brain focuses on waking, sleeping, and dreaming as illustrated in Table 1.1. The continual growth of secular, non-spiritual thinking contributes to an atmosphere more conducive to introspection unthreatened by the contamination of theology. Furthermore, we have a well-developed third person model to inspire, constrain, and instruct first person inquiry. By narrowly bounding our investigations, we are less likely to go astray.

Throughout the first half of the twentieth century the reputation of introspection was so bad that only external behavior was worthy of scientific respect. Consciousness was the 'C-word,' never to be mentioned in scientifically respectable circles. If I am right, it is time for the C-word to come out of the closet and celebrate the rightful place of consciousness as a scientific problem. It is time for conscious thought to serve as a tool to the solution of the brain-mind problem.

- The *second point* concerns the future of psychoanalysis, which for the past century was the self-declared science of subjectivity. My position is decisive, well known and controversial. I believe that psychoanalysis is a failed endeavor that cannot be rescued by a mere political alliance with sympathetic neuroscientists. Freud's views were so authoritative, so unyielding, and, in many cases, so wrong that undue effort must be expended to revise them. In my view, psychoanalysis is a religion which many democratic people tolerate (despite the absence of scientific evidence) but from which science can no more expect enlightenment than can be expected from creationism or intelligent design. In taking this strong stand, I forsake the mantle of toleration worn by my hero, William James.

When William James died in 1912, students of his English colleague, the mystic hypnotist, Frederick Myers, tried to photograph James's soul leaving his body. Needless to say, they failed. Of course, James himself didn't know of this negative result (because he was dead) but had he known (via mediumistic communication across the vale) he would probably have said that the photographic experiment of Myers was inconclusive. But, *pace* James and Myers, I say that there is no good scientific evidence for any parapsychological phenomenon, including the transmigration of souls. Hypnosis is another question which I expect to yield an answer from brain science.

I will focus my case for responsible introspectivism and my case against psychoanalysis on the egregious errors of the foundational propositions in Freud's dream theory. Robert McCarley and I have shown that the disguise-censorship center of Freud's hypothesis was based on misconceptions already apparent in his 1895 *Project for a Scientific Psychology* (Freud 1895). Wisely, Freud abandoned his Project but he unwittingly carried the erroneous neurobiology into his dream theory as erroneous psychology. The goals of Freud's *Project for a Scientific Psychology* are now beginning to be realizable and should be pursued without recourse to the unproductive revisionism of psychoanalysts.

As daunting as it may be, we must admit Freud's mistakes and begin anew. A key part of my appeal will be the necessity of radically revising our concept of the unconscious mind, or as I prefer to call it, the non-conscious brain-mind. When dreaming is regarded as a state of consciousness occurring in sleep instead of an unconscious mental process, the change in emphasis becomes substantial. Non-conscious processes color dreaming consciousness at least as much as they color waking consciousness. Indeed, most of the information in the brain is non-conscious in both states. Some aspects of the non-conscious mind are more available to waking consciousness, while some are more available to dreaming consciousness. Waking and dreaming consciousness are thus reciprocal and interactive, not antagonistic and competitive.

4.2 Mental Illness Is a Misnomer

Nowhere is the necessity for radical change more apparent than in clinical psychology and in psychiatry. Both fields are still held back by the prejudice against introspectionism as well as their sometimes blind faith in psychoanalytic interpretation. By viewing dream consciousness as more akin to the delirium of organic psychosis than to a psychodynamically engendered neurosis, it becomes clear that brain mechanisms determine such forms of subjective experience in dream consciousness as hallucination, delusion, disorientation, single-mindedness and amnesia. These are major symptoms of so-called mental illnesses but, in dreaming, they are normal! In this sense, the interests of Sigmund Freud and Carl Jung in dreaming as informative of unwelcome disturbances in mental functioning during waking appear to be justified (Jung 1957). However, I advocate an important change of emphasis from the purely psychological to the neuropsychological domain.

I now believe that the term 'mental illness' is obsolete and should be discarded. In our hospitals we are dealing with persons afflicted with disorders and diseases of the brain-mind. It is as humbling as it is enlightening to realize that all of those brain-mind features are normal, healthy, and even adaptive when they occur within our sleep.

To appreciate the distinction between "mental" and brain-mind illness, consider the new experimental findings regarding lucid dreams illustrated in Fig. 4.3 By shifting the activation of the frontal brain in the direction of waking, dreamers enter

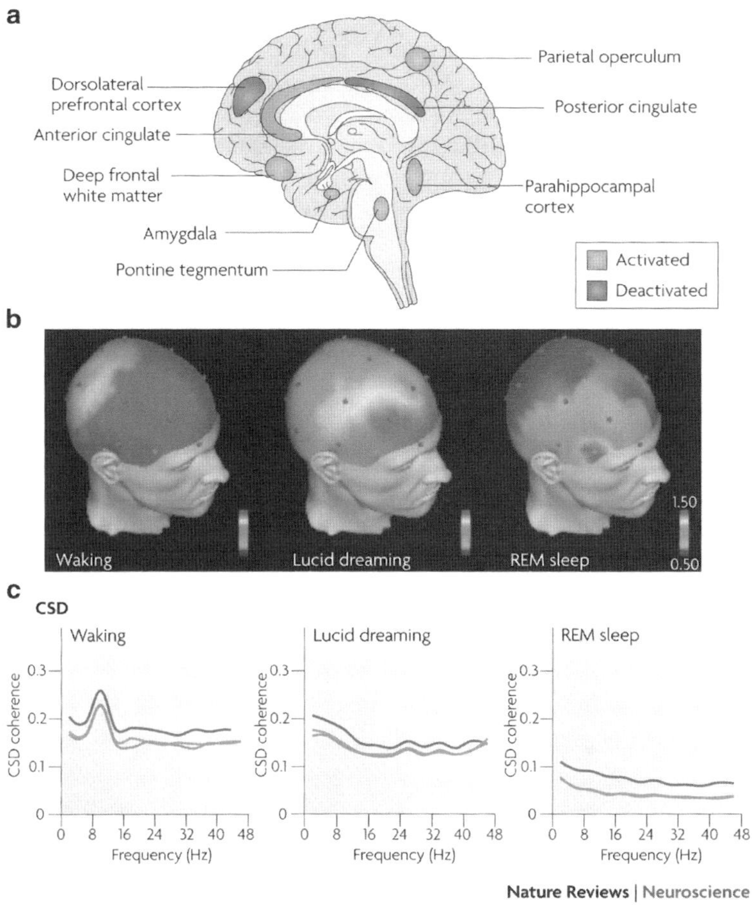

Fig. 4.3 Lucid dreaming. Normal and lucid dreaming: differential regional activation patterns. (**a**) Positron emission tomography studies have measured activation in brain areas during rapid eye movement (REM) sleep in healthy controls and after stroke. Blue and red areas indicate regions showing hyperactivation and deactivation, respectively, during REM (compared with waking) in normal controls. Lesions in two of the activated zones, the parietal operculum and the deep frontal white matter, are associated with reports of cessation of dreaming in subjects following stroke or prefrontal lobotomy. The activation state of the primary visual cortex during REM sleep that is detected in imaging studies is controversial but, contrary to what one might expect, this area is clearly not hyperactivated. (**b**) Quantitative electroencephalographic (qEEG) studies comparing brain activity during waking, lucid dreaming and REM sleep. Frontal areas are highly activated during waking but show deactivation during REM sleep. During lucid dreaming there is an increase in 40 Hz power and coherence in frontal areas compared with non-lucid REM sleep. In lucid dreaming, additional electrical activation of the brain is needed to activate the dreamer's forebrain enough to recognize the true state without causing waking and thus terminating the dream. Differentiated regional activation may underlie the phenomenological distinction between the states of REM sleep, lucid dreaming and waking. Scale bars indicate standardized power based on scale potentials (0.50–1.50 % power). (**c**) In addition to the increased 40 Hz EEG power in frontal EEG leads, EEG coherence (that is, the degree to which the brain waves are synchronized

a unique state of consciousness characterized by both wake like rationality and psychotic like dreaming. Lucid dreaming must be like a "mental illness."

"Dual-aspect monism" may seem to some an unattractive label, especially due to its suggestiveness of compromise–of attempting to have it both ways: the solid substance of materialism and the evanescent gossamer of idealism. I once heard the Nobel laureate neurobiologist Roger Sperry speak of emergentism and liked what he said (Sperry 1981). There is a material thing, the brain, which is organized in such a way that consciousness 'emerges' from its activation. Consciousness is thus a function of the brain. What does the new science of sleep and dreaming have to say about that?

Consciousness does in fact "emerge" as the brain is activated in both waking and in REM sleep. What do these two brain activated conscious states have in common? The occurrence of low-voltage fast activity in the EEG can only occur when the spontaneous oscillation of the thalamocortical system (seen as the spindles and slow waves of NREM sleep) is suppressed. To produce activation, it is enough to drive the system cholinergically to activate the neurons of the nucleus reticularis thalami which inhibit primary thalamocortical projective neurons and stop their spontaneous oscillation so that they can direct information to and from the cortex. According to Giulio Tononi, such information *is* consciousness (Tononi and Edelman 1998) (See again Figs. 3.1, 3.2 and 3.3).

4.3 The Emergence of Consciousness

If you are a human, with an otherwise intact brain, consciousness emerges when your corticothalamic system is activated. This is not the exact meaning of emergentism that Sperry intended, but it is scientifically current. When one is sleepy, for whatever reason, one can feel consciousness ebb gradually away until it is lost altogether. Alertness can be regained by an extreme effort of will and maintained by the voluntary focus of attention, by movement, or by a change in activity.

Falling asleep while reading is so common that many people use reading as if it were a sedative medication. Happily enough, it is not so easy to fall asleep when writing. (If writers fell asleep as easily as readers do, there would be no writing to put readers to sleep!) What is the difference between reading and writing? The answer to that question could help us understand why consciousness is so easily maintained and so easily lost. Two mutually enhancing processes may be at work in writing. One is the motor act of writing which may involve the dorsolateral prefrontal cortex (DLPFC), the seat of so-called executive ego mechanisms. In

---

Fig. 4.3 (continued) across regions) is much higher during lucid dreaming than during non-lucid REM sleep. In fact, the EEG coherence during lucid REM sleep is roughly equal to that during waking (*left panel*), except for the 8–12 Hz alpha range, which shows a peak during waking. CSD, cross-spectral density (Reproduced, with permission, from Hobson 2009) (Color figure online)

order to be a writer, I must focus my attention and perform an extensive range of look-up operations to draw stored information from memory and to synthesize it with the new information which makes my writing as worth reading as it is.

I should do the same thing when I read, but I don't. As in hypnosis, I let the author do the synthesizing for me, and I am happy if I can follow along comprehendingly. It is clear that reading, especially normal fiction and non-fiction, is not as activating a process as is writing. My hypothesis is that the dorsolateral prefrontal cortex plays a key role in keeping me on track in whatever state of consciousness prevails. The fact that the DLPFC is not activated when we dream is a key point supporting the inference that I am kept awake when that part of my brain is doing its job.

So what about dual-aspect monism? How does that messy label help us integrate this data? One major advance is that the brain is not an undifferentiated blob or even a syncytium of 100 billion like-minded neurons. Instead, it is highly differentiated in its parts and powerfully coordinated across its parts. One aspect of coordination is electrical as Wolf Singer has so clearly demonstrated (Singer 2001). For me to be conscious I must be single-minded and clearly focused. That means that there must be massive interconnection between disparate parts of my brain and that many, or more likely most, of my neurons are spending most of their time checking up on each other.

4.4 Binding and Consciousness

It is a fact that brain regions are massively interconnected, and it is also a fact that consciousness is so strongly associated with coordinated activation of the brain's electrical activity (seen as 40–60 Hz oscillations in the low voltage fast EEG) as to suggest that high frequency synchrony of neurons is an organizing factor. Put very simply, the whole brain is in one state (and it is in only one state) most of the time. This is why waking consciousness is not, in William James' famous words, 'a great buzzing confusion.' Dreaming is more like the great buzzing confusion that James eschewed, yet the dreaming brain-mind is still remarkably well organized and unified despite the absence of external space-time and despite deficient neuromodulation.

As already emphasized in Lecture I, Ursula Voss has recently reported that when human subjects become aware that they are dreaming during REM sleep there is a significant increase in the 40 Hz power of their frontal EEG and there is also increased coherence between posterior and anterior brain recording sites (Voss et al. 2009). That so-called "lucid" dreams differ from non-lucid dreams is further evidence that any shift toward waking consciousness is associated with a distinctive change in brain electrophysiology. This leads us to hypothesize that waking consciousness depends upon high frequency brain activation.

If acetylcholine and dopamine working together are enough to produce the thalamocortical activation that we see in both waking and REM sleep, we may wonder if the subtraction of norepinephrine, serotonin and histamine is enough to account for the loss of continuity, the incongruity, and the enfeeblement of all of those cognitive skills that depend on memory? Is demodulation enough to account for the bizarreness, the single-mindedness, and the amnesia of dreaming? Does dream consciousness emerge when the brain is activated in sleep and does it have its unique features because the activation pattern is different in two important ways, one regional and the other chemical?

In order to create a general working hypothesis I will tentatively answer "yes" to all three of these questions. It is already unmistakably clear that the state of the brain and the state of the mind are parallel if not, as I myself hold, two sides of the same coin. It is intriguing to wonder if differential modulation causes both the selective activation of the brain and dream consciousness. An experiment to test the hypothesis that DLPFC activation induces a shift from dreaming to waking is to turn on the DLPFC during REM sleep by simple electrical or transcranial magnetic stimulation (TMS). Or we might give human subjects pilocarpine intravenously to trigger earlier REM sleep onset and dreaming and see if DLPFC stimulation would work against this effect. Whatever the answer to these questions is, it is obvious that we need to learn much more about the DLPFC in regard to consciousness. Sleep and dream research can help us to achieve this goal.

4.5 Is Ego Localizable?

Can working memory, volition and self-reflective awareness be operationally defined? What is the status of these cognitive functions in introspective data? What could be more crucial or central to our interests? And yet, as stressed in Lecture I, no attention has yet been paid to those crucial functions by scientists interested in consciousness. It would seem that there might be a discrete part of our brain, (our self? our ego?) which makes yes/no and go/no go decisions based upon its assessment of the risks and benefits of real or imagined behavioral engagements. The thorny issue of free will lurks around this discussion. But short of tackling this difficult question we can at least assert that while dream consciousness may appear more free (or less constrained) than waking consciousness, it is not at all clear that dreamers 'decide' to do anything or exercise even illusory free will in any way. When, in a dream, I say I am riding a bicycle in downtown London, I do not claim that "I" decided to do so. My brain made that decision for me. I just plug this automatic action element into a story. The synthesis half of the activation-synthesis model is thus enriched.

Self-agency, or responsibility for fictive dream acts, may be intrinsic to REM sleep brain activation or it may be inserted, after the fact, when constructing a dream report. In order to render the report conventional with respect to waking consciousness, an intimation of volition may be introduced, unconsciously, as it were. When it comes to dreams, no assumption of conventionality, elaborated in waking consciousness, can go unquestioned. Folk psychology can be, and often is, simply wrong. Most experienced dream reporters say that they decided nothing at all in their non-lucid dreams. Things just happened to them, spontaneously as it were. They did not have the dream as much as the dream had them!

This paradox strongly implies that the brain has a mind of its own and that the sense of first person agency is entailed by brain activation. In other words, the so-called "self" is an automatic construct. A similarly radical idea has been put forth and elaborately developed by the German philosopher Thomas Metzinger in his books, *Being No One* and *The Ego Tunnel* (Metzinger 2003, 2009).

The occurrence of brain spontaneity and autonomy is very problematical both intellectually and morally. By contrast, the idea of a Newtonian universe with strict deterministic rules is extremely attractive and most people choose to believe in such a model. But the Newtonian world-model may be wrong, or at least badly flawed. If so, then much of our behavior may not only be unwilled, it may not even be caused by identifiable, extrinsic stimulus conditions (*pace* B.F. Skinner). The fact that behavior is sometimes determined by its consequences doesn't mean that it always is.

Dream Vision.
Jay Vogelsong takes inspiration from Odilon Redon whose emblematic eye beholds the dream world.

4.6 Automaticity and Spontaneity

Here we are at the border of philosopher Daniel Dennett's zombie country (Dennett 1991). We do not like to think of ourselves as automatons. But consider for a moment. Suppose we had to remember to go to sleep, or to have 1.5 hours a day of REM sleep? We should be grateful that sleep, including REM, is a built-in, spontaneous and highly reliable brain function. Do we want to be responsible for breathing or not breathing? Both automaticity and spontaneity have their place at the table and are welcome as long as they do not impinge upon functions such as consciousness and its radical darling daughter, free will.

But spontaneity, unpredictability, chaos, and randomness all do creep into psychophysiology as soon as we take an idea like dual-aspect monism seriously. The very words spontaneity, unpredictability, and dissociation are enough to send most people running for the security blanket of dualism. We all yearn for the comforting notion of a divine creator who keeps everything in order via the perfection of his design. Short of that extreme is some sort of idealism, or Platonism, in which the absolute domain of a divine creator, is taken over by the abstract but specific power of ideas. Ideas can be absolute and they can order our thoughts absolutely without the need even of the parallelism that characterizes Cartesian dualism.

I hold that the absolutism of Platonistic ideas is as tyrannical as creationism. In both cases dreamers and wakers are being guided through their life scripts by a power outside their control. In both cases freedom and responsibility are given up in exchange for security and protection. Since we are social beings, we have idealists and dualists praying for us and sometimes believe that our survival is guaranteed by the divine intervention that shapes and directs our behavior despite our apostasy and our willfulness. We are lucky to live in such a tolerant, protective world. If we were powerful kings rather than relatively harmless philosophers, the world would be a sorry place indeed. Even kings wouldn't like it.

4.7 The Hard Problem

Now let's get to work on the hard problem. Imagine a single neuron. Depending upon its genetic design, it is either a pacemaker or a stimulus-follower cell. A pacemaker neuron makes signals whether or not it is told to, but it doesn't make signals if it is told not to do so. A stimulus-follower neuron makes signals according to the constant battle between the excitatory and inhibitory forces impinging on it from other neurons. These two contrasting neuron types are also different in their chemistry: on the one hand, pacemaker neurons are often aminergic; in fact the noradrenergic and serotonergic REM-off cells are pacemakers. They fire unless they are told not to do so.

On the other hand, stimulus-follower cells use other fast-acting neurotransmitters such as glutamate and GABA for their communication with other cells. They do what they are told to do in highly reliable ways. This makes them very good at abstraction or symbolization. Shining a light in the eye results in the obedient encoding of stimulus properties. The code undergoes visual analysis immediately and automatically. Faithful copies are sent into the brain. The codes may be comprised of sequences of signals from one of thousands of stimulus-follower cells. Stimulus information has thus been injected into the system. That information will be combined with internal templates and so create sensation and perception. If the thalamocortical system is activated, if input-output gates are open, and if the system is not occupied with the processing of internal signals, then coded information may pass from the outside world to the thalamocortex for further analysis.

But the brain is not as simple a code reader as a digital camera or a tape recorder. It is, rather, a predictive comparator already equipped with its own expectant codes that are generated, offline, via the spontaneous activation of neurons by internal signals formally similar to those coming from the outside world. The internally generated codes are formally similar because they are elaborated by the spontaneous activity of the same stimulus-follower neurons that are activated by external stimuli in waking. In waking, the intrinsic codes of the thalamocortical system are held at just below threshold by widespread inhibition and/or modulation.

The excitation of an extrinsic code is enough to trigger a response by the intrinsic network. In the case of vision an image is thus created. So far this image is the coded response of intrinsic neurons. But once elaborated, the image can be sent to other parts of the brain for integration with other sense modality codes. I say that once this is done, waking consciousness "emerges." Consciousness is nothing more or less than our subjective experience of the integration of billions of coded neuronal signals. In the case of waking, the visual and related perceptions are a good enough copy of the world to enable the self to see, to evaluate, and to decide what to do about what the self sees. In dreaming, the self also sees well as it produces images in response to internal stimuli. The image generator system may be both disinhibited and demodulated by the withdrawal of aminergic inputs. In other words, as both William Shakespeare and Leonardo da Vinci pointed out four centuries ago, the brain can see in the dark!

4.8 The Seamless Connection Between Brain and Mind

In this view, subjectivity begins with the elaboration of intrinsic codes and builds until it emerges as perception of the world with a self at the center of that world. There is nothing qualitatively tricky about consciousness. As my friend and colleague Giulio Tononi has said, consciousness is nothing more or less than a large amount of information processed in an integrated way.

Of course we need to know much more about how this really works, but I say that we already know a lot more about it than we think we do. Taking sleep and dream science seriously helps and I will try to show you why. When I say that the mind-brain problem may already be solved, this is what I mean. There is no light in the brain. But there is a limitless array of visual image properties encoded in the neuronal assemblies awaiting evocation. These can be used in perception when we are awake to match an external image when we look at an object or a scene, or they can synthesize images if activated by internal stimuli when we dream.

In waking, the ego has not repressed primitive visual functions in order that we may think clearly (as in Freud's secondary process). In dreaming, we do not regress to the visual mode (as in Freud's primary process). When I use Edelman's distinction between primary consciousness and say that we become aware of something like it in our dreams, I am not invoking regression, as did Freud, but rather say that we experience a form of consciousness (Edelman 1992) that is primordial but also foundational. And when I go on to ascribe to waking something like Edelman's secondary consciousness, I mean to emphasize, again, the revelation of the specific brain-mind processes of thought. In this case I move closer to Freud's idea of the ego, but in my use of this concept, I am hoping to explain what the post-Freudian psychoanalysts called the "conflict-free" ego and what modern neuropsychologists call "executive" ego.

According to my protoconsciouness hypothesis, in dreaming we positively activate the visual system and see at the same time that we simultaneously deactivate analytic systems and do *not* think. The purpose of this shift from secondary to primary consciousness is not to protect secondary consciousness. On the contrary, it is to restore the sensitivity and specificity of primary or dream consciousness to support secondary or waking consciousness. This specific example of brain activation shows how close the psychoanalytic and activation-synthesis models can get to one another; but the mechanisms and functions detailed by the two models are diametrically opposite. We are thus forced to choose between one or the other. Psychoanalysis views dreaming as unconscious mental activity that is designed to protect consciousness from disruption by the unconscious mind. Activation-synthesis regards dreaming as evidence of a built-in consciousness generation system.

The scientific study of dream consciousness shows very clearly that the brain-mind can be internally activated even when it is actively cut off from sensory input or motor output. When this occurs, internally generated signals drive the system in such a way as to make us believe that we are seeing. This insight strongly constrains our ideas not only about dream consciousness but also those ideas that we harbor about waking consciousness.

From the subjective evidence of dreaming, we can conclude that formed perceptions, the erroneous assumption that we are awake, and a narrative or scenario structure rich in social, motoric, and emotional content are all synthesized by the brain-mind itself. That capabilities for such an experience are given by brain physiology is also made clear, indicating that the brain-mind, working on its own, creates a model of the world of truly remarkable similitude to the outside world itself. This is primary consciousness.

4.9 Consciousness as Intrinsic Creativity

Instead of merely acting as a predictive model to be checked against sensorimotor reality, the REM sleep activated brain elaborates its own model of the world and spontaneously generates experience-based consciousness. Dream consciousness is thus both partially independent of, and powerfully imitative of, wake state consciousness. Dream consciousness compares formally with waking with respect to its ability to create sensorimotor imagery. With respect to those critical cognitive functions which are in abeyance, it is quite different. Both the perceptual enhancement and the cognitive limitations of dream consciousness have specifically identified physiological bases. This paradigm shift is an enormous advance over the previous state of our science. Now we are even in a position to propose a radically new functional theory–that of dream consciousness as a building block for waking consciousness. That is what I mean when I say that REM sleep dreaming is a protoconscious state.

The existence and nature of dream consciousness reinforces the model of waking consciousness advanced here. It is inconceivable that the brain-mind states of wake and dream consciousness are not somehow complimentary and reciprocal to each other. Thus it is reasonable to assume that, on the one hand, waking consciousness depends upon at least the attributes revealed in dreams, and that dream consciousness, in turn, depends upon the enhancement of some, and the weakening of other, aspects of waking consciousness. Besides complimenting one another, the functioning of consciousness in waking interacts with the functioning of consciousness in dreaming. It remains to be determined exactly how. We already know enough to make intelligent hypotheses of what this functional reciprocity might be. The REM effects on waking include the enhancement and revision of learning and memory (see Fig. 4.4), the refreshment and maintenance of temperature control networks, and the balance of circuits necessary to assure psychic equilibrium.

These surprising functional benefits outweigh and challenge the psychoanalytic idea that dreaming is the guardian of sleep. Besides being philosophically untenable (since psychological events do not, as far as we know, *cause* physiological responses), Freud's sleep guardian theory is replaced by the physiological processes that guarantee the preservation of sleep in the face of internal brain activation. We also know, from numerous empirical studies, that sleep enhances memory (Marshall and Born 2007; Siegel 2001, 2005; Stickgold et al. 2000; Vertes 2004; Walker et al. 2002, 2003). Some of that evidence, suggesting that the REM sleep that determines our wildest dreams also consolidates procedural learning, is illustrated in Fig. 4.4. It is ironic that a state of the brain-mind whose conscious experience is so difficult to remember should be beneficial to memory. Apparently we need not perceive the operation of this important function in order to be the beneficiary of it. In other words we are again struck by the mutual advantages of waking and dreaming as well as realizing that sleep needs no guardian to persist in the face of brain activation because inputs and outputs are actively blocked by robust physiological mechanisms.

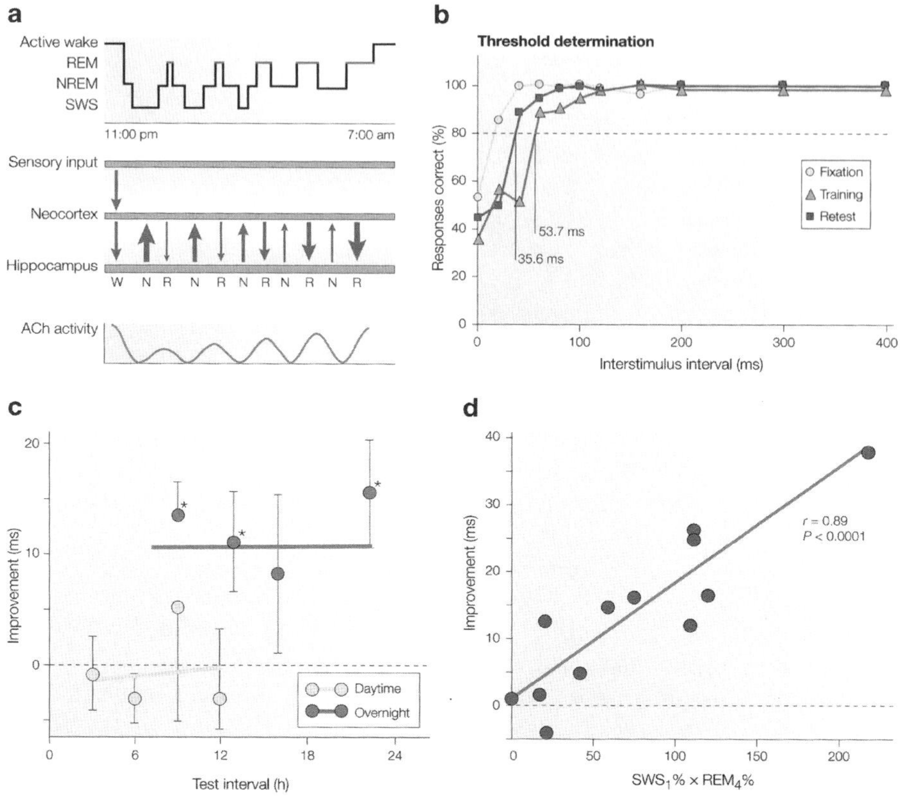

Fig. 4.4 A model of sleep dependent memory consolidation. Supportive results for the sleep-learning function are from the texture-discrimination task (TDT) of Karni and Sagi. (**a**) State-related changes in hippocampal-neocortical information flow. Changes in cholinergic (acetylcholine, ACh) neuromodulation (*bottom*) and hippocampal-neocortical communication (*middle*) are aligned with the human rapid eye movement (REM)-non-REM (NREM) cycle (*top*). Cholinergic levels are maximal in waking and REM, and minimal in NREM. During waking, environmental input to the primary sensory neocortices proceeds through association cortices to the entorhinal cortex and into the hippocampus. Reverse flow from the hippocampus to the cortex is attenuated. During NREM, information is conveyed primarily from the hippocampus to the neocortex, where, over time, the memories that this information represents are permanently stored. During REM, as in waking, this hippocampal outflow to the neocortex is blocked, but wake-like flow of information from neocortex to hippocampus might again be possible. Changing levels of acetylcholine favor this pattern of state dependent information flow by augmenting feedforward (cortical-hippocampal) transmission circuits in the hippocampal formation, and by blocking feedback (hippocampal-cortical) circuits during cholinergic maxima in wake and REM. During cholinergic minima in NREM, such feedback circuits are released to allow flow of information from the hippocampus to the cortex. *N* non-REM, *R* REM, *W* waking. (**b**)The Karni and Sagi TDT. Subjects view a textured pattern that is composed of short lines on a computer screen while keeping their vision focused on a central fixation point. In each trial, they are briefly presented with a stimulus to detect at the fixation point, while being asked to decide whether three short lines in one quadrant of their peripheral visual field are aligned in a vertical or horizontal pattern. After an interstimulus interval (ISI), they are presented with a masking computer screen that is composed of multiple lines that effectively extinguish any afterimage of the stimulus screen. The ISI is then

4.10 Philosophy of Psychology

The activation-synthesis model of dreaming has important points of difference, but also important points in common, with existing psychological theories. It is most easily incorporated into cognitive behavioral theory and may even form a solid base in neuroscience for that theory. It is less easily reconcilable with Freudian psychoanalytic theory and its orthodox psychodynamic derivations. However, what is now taken as psychoanalysis has little substantive resemblance to the orthodox position of Freud, so it may even be useful in forming a biological base for a non-orthodox version of psychodynamic psychology.

The activation-synthesis model is very compatible with the concept of the cognitive unconscious formulated by the American psychologist, John Kihlstrom (Kihlstrom 1987). It shares with Kihlstrom's ideas the recognition that most of the information in the brain-mind is non-conscious but that, contrary to Freud's view of a dynamically repressed unconscious, much of the memory content in the unconscious is accessible to waking consciousness. Dream consciousness is seen not as a purposefully distorted rendition of dynamically repressed unconscious wishes but rather as a direct and undisturbed read-out of non-conscious and conscious mental content. Note that this difference is better appreciated by the adjective "non-conscious" because the word "unconscious" comes with so much Freudian baggage that it can hardly be used without conjuring up images of repression.

To make this point indelibly clear, it may be worth comparing the activation-synthesis hypothesis of dreams with orthodox psychoanalytic dream theory (PDT). According to PDT, dreaming occurs because unconscious infantile wishes, which are easily suppressed during waking, become active in sleep. When the ego is off duty, the id becomes unruly. To the rescue of the sleeping ego come the defensive

Fig. 4.4 (continued) progressively shortened over successive trials of a training or testing block (see *below*). The dependent variable, derived by interpolation, is the minimal ISI at which they are 80 % correct in their decision on the vertical-versus-horizontal alignment of the bars (53.7 ms in this figure). Subjects first complete a 60–75 min training session, and are then retested in an identical testing session after a period of time in which experimental manipulations (varying duration, sleep deprivation) can be performed. (**c**) Time course of improvement on the TDT. Subjects were tested either on the day of training with no intervening sleep (*yellow circles*) or on the day after training following a night's sleep (*blue circles*). Only subjects who slept for 6 or more hours between the training and testing sessions showed improvement in TDT performance. Asterisks indicate individual groups showing significant improvement. (**d**) A two-step model of memory consolidation in TDT performance. Improvement in TDT performance was found to correlate with the percentage of slow-wave sleep in the first quarter of the night (SWS1) and the percentage of REM sleep in the last quarter (REM4), but not with sleep-stage variables from other parts of the night. Most notably, the product of percentage SWS1 and REM4 further improved the correlation with TDT improvement when compared with either measure individually, indicating a two-stage process for the consolidation of improvement on this task. (Summary of data and concepts put forward by Robert Stickgold when he was working in my laboratory) (Reproduced, with permission, from Hobson 2009) (Color figure online)

forces of disguise and censorship. They bowdlerize the kinky id forces and make them look nonsensical and meaningless whereas, in fact, they are masquerades for viciously potent entities that would overwhelm consciousness if admitted to that realm undisguised.

Continuity theory is a related offspring of modern sleep and dream research. Espoused at first by such psychologists as David Foulkes and John Antrobus and more recently championed by William Domhoff and Michael Schredl, continuity theory emphasizes the similarities between waking and dreaming consciousness (Domhoff 2002; Domhoff et al. 2006; Foulkes 1962; Singer and Antrobus 1965; Singer and Antrobus 1972). Protoconsciousness theory is in accord with this aspect of continuity theory. There is no doubt that waking and dreaming are similar but why they are similar, and, more significantly, why they are so different, separates the two schools almost completely.

From the beginning, the continuity school held that waking and dreaming were only different because the brain activation of REM was less intense than that of waking. While the activation level distinction has been upheld by physiology (and is quantified as Factor A), physiology has added input-output gating and modulation to the mix. These two parts, Factors I and M, complete the three dimensional structure of the AIM model and distinguish that model from the activation-only model of continuity theorists. Activation-synthesis holds that these two additional variables *must* determine such psychological aspects of dreaming as the hallucinosis (Factor I) and the amnesia (Factor M), two robust formal features of dreaming that distinguish it from waking in ways unexplained by the activation-only model.

It may be no coincidence that continuity theory (which I hold to be correct as far as it goes but I claim it does not go far enough) tends to be accepted by psychologists who are either ignorant of, or downright hostile to, physiology. Like the psychoanalysts they seek to distance themselves from, these researchers would prefer to think that physiology has nothing much to offer dream theory. They seem to want to possess dream theory as a purely psychological process in order to protect it as intellectual territory. Nothing could be further from the integrationist spirit of William James, a spirit now even more important to psychology than it was in 1890 (James 1981).

4.11 This Mimic Wakes

In 1727, John Dryden wrote:

> Dreams are but interludes, which fancy makes
> When monarch reason sleeps, this mimic wakes
> Compounds a medley of disjointed things
> A mob of cobblers and a court of kings.

Agreeing with Dryden (Dryden 1970), activation-synthesis has a radically different view of the process from that of Freud and the Continuity School. So different, in fact, that it is not as much a matter of revising those theories as replacing them. Dream instigation is the automatic psychological concomitant of brain activation in sleep. Most of the material of which dreams are composed is non-conscious, but it is *not* dynamically repressed and it is *not* a replay of waking experience. Endogenous visual images are suppressed during waking to prevent hallucinations, which would be quite inconvenient. Sub-threshold images may actively aid in image formation, where image formation, in waking, is recognized to be in the service of accurate stimulus recognition and in the creation of a reliable model of the outside world.

This view completely reverses the notion of the unconscious as the enemy (or jealous rival) of the conscious mind, and says, instead, that unconscious forces are trying to help, not hinder, reality testing. This is a point that cognitive behavior theorists will welcome, if they have not already discovered it for themselves. For continuity theorists, the operation of an endogenous imitator of waking may be easily accommodated but, if, and only if, the continuity arrow flies in both directions, into dreaming out of waking and into waking out of dreaming. This is the only continuity that fits with the data.

Dynamically repressed (or actively forced down) mental content may well emerge in the process of dream image creation and plot selection processes that activation-synthesis credits with dream production, but such material is neither necessary nor sufficient for dreaming to occur in sleep. Dreaming is neither the only, nor necessarily the most privileged, way of getting at those psychic residues of trauma and conflict that constitute a kind of informational infection of the brain-mind.

For the continuity theorist, it is important to acknowledge, as emphasized above, that continuity is a two-way street and that waking and dreaming are as significantly different as they are similar. The extreme position of some continuity psychologists, that waking and dreaming are identical, is now clearly untenable in the face of empirical studies of dream phenomenology summarized in Table 2.1 and the physiological distinctions between REM sleep and waking outlined in Lecture II.

4.12 Emotion and Consciousness

Most dream content is unconflicted despite its association with the strongly predominant emotions anxiety, elation, and anger. The dominant presence of these three emotions in dreams is a fact that cannot be explained by psychoanalytic dream theory, but it is easily and comfortably dealt with by the activation-synthesis hypothesis. Furthermore, it is not, in itself, at odds with continuity theory.

During REM sleep, brain emotion centers in the amygdala, the frontal lobes, and the parahippocampal cortex are activated in parallel with activation of sensorimotor and corticothalamic circuits. There is strong physiological evidence for both activation pathways. For the activation-synthesis hypothesis, the evocation of these three cardinal survival emotions is altogether welcome.

For psychoanalytic dream theory, emotion is a problem as Freud made clear in his 1933 "revision" of his 1900 theory (Freud 2001). Since anxiety was regarded by Freud as a neurotic symptom, it was difficult for him to explain why it should be so prominent if dream disguise and censorship were working effectively. Freud never came to terms with the fact of dream anxiety. He should have recognized, on emotional grounds alone, that his dream theory was badly in error.

Protoconsciousness theory regards anxiety as an important attribute of consciousness. On Darwinian grounds, it makes sense for organisms to be wary, to be skittish, and to be ready to run away. If they did not flee, they might quickly be someone else's lunch! Anxiety can, when exaggerated or amplified, become a symptom and it may even require treatment. But anxiety aids survival. It is thus existential and a strong guide to waking consciousness. Its presence in dreams is testimony to its importance in waking.

Continuity theory (and psychoanalysis) are both at pains to explain the surprising under-representation in dreams of the so-called social emotions (shame, sadness, and guilt) whereas activation-synthesis sees these emotions as dependent upon social context and, as such, functions of the secondary consciousness processing of the waking state.

The bizarreness of dreams, which psychoanalytic dream theory attributes to disguise of the troublesome unconscious wishes, is viewed by the activation-synthesis hypothesis as the understandable by-product of two non-conflictual factors: (1) the unavailability of the real world space-time continuum and (2) the chaotic nature of the REM sleep brain activation process. Critics of the activation-synthesis hypothesis are troubled by the notion of chaos and are completely undone by the idea of randomness. For the strict Newtonian determinist (and Freud was certainly one of them), there is no such thing as an undetermined (or chaotic) event. Now, over 100 years later, quantum physics, chaos theory and the uncertainty principle are all enshrined as basic tenets of physics and chance is seen as an essential ingredient of all natural processes (Hobson and Friston 2012, 2014).

The brain is a physical object of such enormous complexity as to boggle the mind. But a moment's reflection helps us to see that chaos and randomness are our allies, not our enemies. Elsewhere I have expanded upon the idea that without chaos and without randomness, we are without freedom or, at least, the comforting illusion thereof. Here it suffices to say that the unpredictability that goes hand in hand with chaos, is the friend of creativity and novelty; it is also bail money for release from the otherwise inescapable jail of the repetition compulsion.

Unpredictability is the friend of dream scientists who cannot explain why a given dream plot was chosen on a given night. Given the difficulty that dreamers have with the identification of memory sources for dream plot items it seems likely that specific dream plot selection may never be explained in a historical way. Plot selection is commonly seen as a result of something that happened in waking. But since dreams are not mere replays of waking events, we may have to wait for a better theory to account for dream plot selection. For now, we cannot say why a given dream occurred on a given day. Dream content, as Douglas Hofstadter has emphasized (Hofstadter 1999), is almost completely unpredictable. Because REM sleep dreaming is redundant and profuse, it be better to be able to touch all the internal bases than to activate a system that is driven exclusively by external data. In other words, dream consciousness safeguards its own comprehensiveness via automaticity and chance.

4.13 Dream Plot Selection: Another Look

Psychoanalytic dream theory was compounded of two dream causation factors. One was the unconscious wish, and I have already expressed grave doubt about that factor. The other was the so-called day residue, which Freud said paired up with the unconscious wish to cause the dream. It is true that current events do crop up in our dreams, but experiments show that they do not usually appear on the night of their daytime occurrence. The peak date of current event incorporation into dreams is 6 days earlier. So the Freudian day residue hypothesis must be modified accordingly.

The fact that dream plot selection and dream content itself cannot be explained by a simple rule of recency or potency is accepted by the activation-synthesis hypothesis and welcomed as a guarantor of the thoroughness of whatever information generation process is at work. Any and all non-conscious information is a candidate for plot selection and appearance in dream scenarios as long as it is compatible with dream animation. Remember the word creative, implying imagination and even artistry. Here again the activation-synthesis hypothesis is happy to applaud the lawlessness of dream construction because it takes that feature to be important to a view of the brain-mind as a constant novelty seeker and novelty creator. The brain-mind is not a plodding and repetitive automaton destined, forever, to loop around safe but uninteresting cerebral and ideational circuits.

Why are dreams forgotten? That is another tough question for psychoanalytic dream theory. Are they re-repressed? If they have been bowdlerized, why do they have to be forgotten? Activation-synthesis hypothesis notes the aminergic demodulation to 50 % of waking levels during NREM sleep and 100 % during REM sleep, and says, simply, that memory is disenabled during sleep. You are not supposed to remember your dreams. You must just appreciate the benefits of REM sleep. By all means, avoid the trap of the psychoanalyst's couch (which you can't get off of without dream recall–and the right dream recall at that). I will not even mention the money you will pay your analyst as you look for the meaning of your life in your dream tea leaves.

4.14 Psychotherapeutic Efficacy

This brings us to the theory of psychotherapeutic efficacy. Cognitive Behavioral Therapy (or CBT) would say it hopes that clients are helped to understand themselves better and to change their behavior so that it is more adaptive. This goal may be facilitated by talking with a sympathetic, neutral and experienced observer. I agree. It may help. Whether it is helpful enough for an insurance company to underwrite the process as medical treatment is another matter, far beyond the scope of this discussion. Dreams are a part of life, a zesty part, and an informative part so why not discuss them, too, as part of CBT? Sometimes dreams do reveal that earlier life issues, long believed dead, are still very hot in our non-conscious brain-minds. Freud does deserve credit for insisting on the long-term persistence of conflict and trauma.

Unjustified by dream science, however, is any interpretation scheme based upon the central Freudian hypothesis that forbidden infantile wishes cause dreams via the stimulation of disguise-censorship mechanisms. The idea that such wishes (the latent content of the dream) can be uncovered and identified confidently by free association to the remembered (or manifest) content of the dream is without any scientific support. And whether such a process would benefit an individual's mastery of life is also undocumented. There is not even solid evidence to suggest that any discussion of dreams is either necessary or useful. Dream discussion could be an interesting but empty exercise if the goal is simply to change behavior.

If one of our goals is to understand behavior, then paying attention to dreaming may be quite useful as I hope I have already made clear. Freud said that dreaming was the royal road to the unconscious. Little did he know that it was consciousness itself that was crying out for understanding via the scientific study of dreams. Freud made this mistake because his outmoded neurology led him to believe that the unconscious was a repository for unrealized impulses and ideas that were the enemy of consciousness.

Viewing the unconscious brain-mind as an ally of consciousness changes everything. Instead of a layered system with the unconscious as the lowest layer (along with base instinctual impulses), we conceive of a parallel system in which the non-conscious brain plays a vital role. It is not seen as predominantly (if at all) a repository for unwanted instinctual issues but rather as a treasure chest of supportive allies to the analytic, executive, and creative functions of the brain mind.

4.15 Confabulation as One Ally of Consciousness

How might such a system work? Let us go back to our consideration of a visual stimulus entering the brain. Let the stimulus be my collaborator's hat on the rack outside my office. Within milliseconds, a formal analysis of the stimulus is complete and the fruits of that analysis are fed to the temporal lobe for recognition. What is it? A hat is the answer. Hats are worn by people (who tend to be about the same height as the rack on which the hat sits). Suddenly I see my collaborator with his hat on his head waiting to leave the lab with me. I am surprised, because I did not know we had a plan to leave together. I look more closely at the hat and my co-worker disappears. His hat is still on the rack outside my office door. It all happened so quickly that I almost didn't notice this classic example of 'filling in.'

Filling-in is the brain-mind's way to complete an expectation. But, as surely as if I had a REM sleep dream, complete with wild visual hallucinations, I have just had, in waking, a brief, subliminal hallucination.

My brain has supplied itself with the image of a person to wear the hat. This is activation-synthesis with a vengeance but now it causes hallucination in waking. That waking hallucination is not sustained because Bob McCarley is not really standing there, under his hat. Only the coat rack is. I am not, thankfully, afflicted with major mental illness. So I rapidly throw out the image hypothesis that Bob is standing there and go on about my own wake state business, which is simply to go home. Before leaving this trivial, but telling, anecdote, let me say that I would estimate the total time elapsed in this process to be less than 1 s.

How can the system work that fast, you might well ask? In REM sleep our PGO wave candidates for pseudo-visual stimuli occur at a rate of 6–8/s. That gives 125–150 ms for each wave. 125 milliseconds is one eighth of a second. That is fast. Even if PGO waves are only timing signals, they provide an important temporal frame for perceptual analysis to sketch in. In dreams, as long as the broad outlines of a plot are not violated, the characters, the locus, and the action of a dream scene are not changed. The man in a hat may be identified as Bob even though the dream image man does not resemble Bob. The name, Bob, matches the hat. That is good enough for starters. I guess it is Bob, so it is Bob. My critical faculties are too weak to reject that erroneous hypothesis, and it is so weak that it accepts other still more flagrant disorientations.

4.16 Machine Analogies and the Brain-Mind

Our tendency to use tape recorder, computer, or photographic camera analogies for the brain do not make it easy for us to imagine that perceptual image construction and implanting might occur six to eight times per second (or more) but of course the speed of conduction in the larger nerve cells is prodigiously fast, with rates of up to 100 m/s. The world record for the 100 meter dash was broken in Beijing this year by the Jamaican runner, Usain Bolt. He covered the distance in 9.4 s, almost ten times slower than the conduction speed of his own motorneurons. At that rate, a brain stem signal can reach the cortex (a distance of 0.1 m) in a few milliseconds. Add 500 ms (this is ½ s) for each synaptic gap crossed and you are still operating in a high speed range (and the system can compensate for delays).

I want to make clear the central idea of this new protoconsciousness theory. It is that the brain constructs an *a priori* model of the world. Immanuel Kant would like my theory (Kant 1999); John Locke would not (Locke 2009). In waking, the brain-mind holds at just below threshold countless stimulus identifier circuits and countless more stimulus association circuits. Association has been the basic law of psychology, and now we see how association really works. In relation to image generation, it is 'filling in.' Beyond that, it is scenario construction. Scenario construction is a synthetic process akin to story telling, to filmmaking and to scientific hypothesis testing all thrown into one conscious representation. It is easy, once this basic principle is grasped, to add in emotive coloring (or perhaps to fill in a plot to fit an emotion).

In any case it should be clear to you by now that I think dreaming reveals many fundamental aspects of how consciousness is created by the brain and why dreaming, however poorly remembered, should not be regarded as an unconscious function but rather as a primordial (or protoconscious) state. It is this central hypothesis which must be integrated into dynamic psychology whatever its therapeutic method used by clinical practitioners might be. Indeed, the view taken here of dreaming as an ally of the waking brain-mind should be seen as a key building block in any general theory of how the brain-mind works, including how it can be helped to work better.

4.17 Activation-Synthesis, AIM, and Clinical Psychiatry

If you want to learn about mental illness at first hand, just go to sleep and have a REM sleep dream. Then, without moving, wake up as completely and quickly as possible. Recall the details of your dream and recognize that dreaming formally evinces many major symptoms of mental illness. In REM sleep dreams, you regularly hallucinate. You see things that are not there; and you are delusionally convinced that you are awake. When dreaming, you are therefore psychotic, by definition. You have hallucinations and delusions. It is faint comfort to regard these processes as non-conscious as if that designation helped you escape the obvious: the hardware for psychosis is built into the brain.

The reason that the hardware for psychosis is built into the brain is not to guarantee our capacity to go crazy every night of our lives. It is a side effect of endowing consciousness with the special skills that I have already enumerated. The ability to elaborate upon visual stimuli, to fill in, to flesh out, and to imagine the impossible, to create, to be original, and yes, to be free. This naturalization of the psychotic features of dreaming has far-reaching impact upon our concept of ourselves, upon consciousness science, and upon psychiatry.

At first glance, this vision of the brain mind as intrinsically prone to psychosis is more frightening than Sigmund Freud's view of the unconscious as a chamber of horrors. Freud's metaphor was more benign in that he saw dreaming as a natural neurosis. For Freud, our unconscious mind was a bag full of nasty impulses which constantly threatened to escape and overwhelm us. Viewing the dream as a natural psychosis is even more frightening. It means that we could all become psychotic by quite natural means. We don't have to have a special gene, or a particular kind of mother, we can just nod off and, whammo, we are all hit with a state of mind formally akin to madness.

4.18 Curing Dream Psychosis

Let's look on the bright side: the psychosis of dreaming is as easily cured as it is induced. All you have to do is wake up! The psychosis that is dreaming does not crawl into a hole and try to escape all day long as in Freud's model. The psychotic state is instead gone and its residual force, now mostly contained, is in the service of reliable perception and justified belief. In a word, activation-synthesis regards even our dream psychosis as potentially useful.

Many critics object to the suggestion that any normal process, like dreaming, can be in any way psychopathological. I am not saying that dreaming is sick, or even abnormal; I nonetheless do hold the view that the same brain-mind functions that support mental health, support mental illness and we would do well to accept that fact and find out why. In that sense, I too am a continuity theorist. For me, the recognition that dreaming is a psychosis is not so much a threat as it is a gift. Understanding the brain basis of dreaming is not just a way at getting a fix on normal consciousness. It is also a way of getting a fix on mental illness.

So what kind of psychosis is dreaming? To answer that question we need to ask a few more questions. What kind of hallucinations occur in dreams? Mostly visuomotor is the answer. Now while visual hallucinations are relatively rare in the so-called functional psychoses, schizophrenia and manic-depressive disorder (now called bipolar), they are very common in so-called organic psychosis. It is true that in organic psychosis, the visual hallucinations may be stereotyped in a way that is never seen in dreams. Nonetheless the visual hallucinations favor an organic basis, and we already know that there is one: it is pulsatile excitation of an aminergically demodulated thalamocortical visual system.

4.19 Functional vs. Organic States: A False Dilemma

Dreaming is a *functional* psychosis if there ever was one. But it is organically determined. So the distinction between functional and organic psychosis breaks down when we consider dreaming. Using the term "functional" for schizophrenic and manic depressive psychoses only means that the organic mediation of those illnesses has not yet been defined. Obviously, everything that the brain does is both organic and functional. For a century, however, psychiatry has misled us into believing that entirely and exclusively functional disorders existed. This myth was in the service of the delusional conceit that mental illness could be treated by exclusively functional means, i.e. by talking. When I was in training, psychotherapy was held by many to be both necessary and sufficient to treat schizophrenia. Time has since shown this claim to be false.

The delusions of dreams are never paranoid as so often are the delusions of schizophrenia and manic-depressive psychosis. Although affirmative probe studies have not been done to check out this supposition, such delusions as persecution by the FBI, sacrifice by God, and other outlandish claims, do not normally appear in dream reports. And subjects only rarely report that they become aware that they are not really awake. Lucid dreaming is uncommon but it can be induced by pre-sleep autosuggestion. This means that there are, indeed, dynamic interactions between waking and dream consciousness.

The lack of insight that is so characteristic of dreaming is striking given the panoply of discontinuities and incongruities that constitute dream bizarreness. You would think that ambiguous identities of dream characters, incongruous details in dream scenes, and even impossible physical actions like flying, levitating or skimming across water would tip us dreamers off, but they usually do not. Viewed as microscopic disorientation, the instability of time, place and person that is characteristic of dream thought is typically seen in one and only one kind of psychosis, the organic type where it is typically associated with two other dream features, recent memory loss and its double, confabulation. This completes the tetralogy of symptoms and signs of organic psychosis:

1. Visual hallucinosis
2. Disorientation
3. Recent memory loss
4. Confabulation

Without recent memory it is impossible to take account of dream bizarreness and to recognize it as such. Instead of stopping the action and saying: 'Hey, wait a minute, no person has the features of two of my friends,' or 'I can't really be flying,' we sail along in our dream throwing good cognitive money after bad. When dreaming, we are, I submit, organically and functionally psychotic.

4.20 Dream Consciousness and Its Causation

While our emotion when dreaming may be ecstatic–as in mania–we are more often anxious or angry. The giddy high of mania is only occasional. I love my manic dreams as much as I enjoy similarly ecstatic mood states in waking but they are sadly uncommon in both my waking and my dreaming consciousness. But I am spared the embarrassment and cost of buying sprees in both dreams and in my waking life. Surprising is the almost total absence from dreams of social affects like sadness, loneliness, helplessness, and apathy–emotions that are quite common in depression. The dreams of depressed people may be depressed but the dreams of normal people are not. This is surprising for two reasons. One is that some depressed affect is not unusual in normal waking. The other is that the organic determinants of REM sleep dreams are very similar to those of depression.

Aminergic weakness and cholinergic strength are the twin physiological hallmarks of both dreaming and depression. Before speculating about why dream affect is not depressed, I hasten to point out that organic psychosis is also caused by the administration (and especially the withdrawal) of drugs that have profound effects on aminergic and cholinergic neuromodulation. These drugs include the amphetamines, cocaine, and alcohol but also comprise such medically dispensed items as atropine, anti-psychotic 'tranquilizers,' and anti-depressant pills. The good news is that these drugs are both potent and useful in setting the balance of state control mechanisms in the brain. As such, however, the chance for mischief and unwanted side effects, including psychosis, is great. In psychiatry, as elsewhere in life, there is no free lunch.

4.21 Phenomenological and Intellectual Awakening

If memory systems suffer from long-term drug use or from the temporary chemical imbalance of REM sleep dreaming, it is clear that insight and judgment will suffer, too. Thus I am often unable to say that I am dreaming when I am dreaming even though that insight is immediately apparent when I wake up. What has changed? When I awaken, the aminergic waking systems have been turned back on and the cholinergic dreaming systems are throttled. Is it really all that simple? Probably not, but possibly yes! It is something like that simple.

Given the extent of our knowledge it is even possible to assert that waking up is a kind of experiment of nature that invites closer introspection. It is a way of defining and measuring changes in consciousness as one passes, quickly, from dreaming to waking. My reframing of dreaming as an altered state of consciousness rather than an unconscious mental state inspires this approach to subjective data collection. After all it is most likely that it is we humans (and not cats or rats or even monkeys) who enjoy secondary consciousness and stand to gain from its exploration and manipulation.

In reaching this conclusion, I follow Donald Griffin and Gerald Edelman who allowed that subhuman animals may have primary consciousness (where primary consciousness implies perception and emotion), but not secondary consciousness (where secondary consciousness implies awareness of awareness and propositional thought, both of which appear to depend on the evolution of language). Among other skills, we do have language and can give verbal reports of our states. With training we can expect to become better at self-observation in the service of exploring and understanding our consciousness. Sub-human mammals are more likely stuck with something like protoconsciousness. In other words, humans are more likely than other mammals to experience enormous differences between waking and dreaming consciousness.

My own recipe for further work in this exciting field of science specifies a variety of state changes that can be objectified and tracked subjectively by trained subjects (Hobson 2014). States of interest include lucid dreaming, hypnosis, and meditation. Objective methods include quantitative EEG, MEG, scanning for measurement purposes, and electrical and transcranial magnetic stimulation (TMS) for state change induction. I may be wrong when I say that the mind-brain problem is essentially solved, but I know I am right when I say that we have crossed a scientific threshold in consciousness studies. By utilizing the concepts and methods of sleep and dream science, the scientific study of consciousness gains the important advantage of empirical substance as well as promising the integration of two fields that are really one.

References

Chalmers, D. J. (1996). *The conscious mind: In search of a fundamental theory*. New York: Oxford University Press.

Dennett, D. C. (1991). *Consciousness explained*. Boston: Little, Brown, and Co.

Domhoff, G. W. (2002). *The scientific study of dreams: Neural networks, cognitive development and content analysis*. Washington, D.C.: American Psychological Association Press.

Domhoff, G. W., Meyer-Gomes, K., & Schredl, M. (2006). Dreams as the expression of conceptions and concerns: A comparison of German and American college students. *Imagination, Cognition and Personality, 25*(3), 269–282.

Dryden, J. (1970). *Poems and fables of John Dryden*. Oxford: Oxford University Press.

Edelman, G. M. (1992). *Bright air, brilliant fire: On the matter of the mind*. New York: Basic Books.

Foulkes, W. D. (1962). Dream reports from different stages of sleep. *Journal of Abnormal and Social Psychology, 65*, 14–25.

Freud, S. (1895). *Project for a scientific psychology*. New York: Standard Edition.

Freud, S. (2001). New introductory lectures on psycho-analysis. In *The standard edition of the complete psychological works of Sigmund Freud* (Vol. XXII, 1932–1936). London: Vintage Classics.

Genzel, L., Dresler, M., Wehrle, R., Grözinger, M., & Steiger, A. (2009). Slow wave sleep and REM sleep awakenings do not affect sleep dependent memory consolidation. *Sleep, 32*, 302–310.

Hobson, J. A. (2009). REM sleep and dreaming: Towards a theory of protoconsciousness. *Nature Reviews Neuroscience, 10*, 803–814.

Hobson, J. A. (2011). *Dream life: An experimental memoir*. Cambridge, MA: MIT Press.

Hobson, J. A. (2014). *Ego damage and repair*. London: Karnac Press.

Hobson, J. A., & Friston, K. J. (2012). Waking and dreaming consciousness: Neurobiological and functional considerations. *Progress in Neurobiology, 98*, 82–98.

Hobson, J. A., & Friston, K. J. (2014). Consciousness, dreams, and inference: The cartesian theatre revisited. *Journal of Consciousness Studies, 21*(1–2), 6–32.

Hofstadter, D. R. (1999). *Godel, Escher, Bach: An eternal golden braid*. New York: Basic Books.

James, W. (1981). *The principles of psychology*. Cambridge, MA: Harvard University Press.

Jung, C. G. (1957). *The undiscovered self (present and future)* (1959 ed.). New York: American Library.

Kant, I. (1999). In P. Guyer & A. Wood (Eds.), *Critique of pure reason*. Cambridge: Cambridge University Press.

Kihlstrom, J. F. (1987). The cognitive unconscious. *Science, 237*(4821), 1445–1452.

Locke, J. (2009). *An essay concerning human understanding (complete and unabridged)*. Milwaukee: WLC Books.

Marshall, L., & Born, J. (2007). The contribution of sleep to hippocampus-dependent memory consolidation. *Trends in Cognitive Science, 11*(10), 442–450.

Metzinger, T. (2003). *Being no one: The self-model theory of subjectivity*. Cambridge, MA: MIT Press.

Metzinger, T. (2009). *The ego tunnel: The science of the mind and the myth of the self*. New York: Basic Books.

Siegel, J. M. (2001). The REM sleep-memory consolidation hypothesis. *Science, 294*(5544), 1058–1063.

Siegel, J. M. (2005). Clues to the functions of mammalian sleep. *Nature Insight, 437*(27), 1264–1271.

Singer, W. (2001). Consciousness and the binding problem. *Annals of the New York Academy of Sciences, 929*, 123–146.

Singer, J. L., & Antrobus, J. S. (1965). Eye movements during fantasies: Imagining and suppressing fantasies. *Archives of General Psychiatry, 12*, 71–76.

Singer, J. L., & Antrobus, J. S. (1972). Dimensions of daydreaming: A factor analysis of imaginal processes and personality scales. In P. Sheehan (Ed.), *The function and nature of imagery*. New York: Academic.

Sperry, R. W. (1981, December 8). Some effects of disconnecting the cerebral hemispheres. *Nobel lecture*. Accessed 8 Dec 1981.

Stickgold, R., James, L., & Hobson, J. A. (2000). Visual discrimination learning requires sleep after training. *Nature Neuroscience, 3*(12), 1237–1238.

Tononi, G. Psi. Unpublished manuscript.

Tononi, G., & Edelman, G. M. (1998). Consciousness and complexity. *Science, 282*, 1846–1851.

Vertes, R. P. (2004). Memory consolidation in sleep; Dream or reality. *Neuron, 44*(1), 135–148.

Voss, U., Holzmann, R., Tuin, I., & Hobson, J. A. (2009). Lucid dreaming: A state of consciousness with features of both waking and non-lucid dreaming. *Sleep, 32*(9), 1191–1200.

Walker, M. P., Brakefield, T., Morgan, A., Hobson, J. A., & Stickgold, R. (2002). Practice with sleep makes perfect: Sleep-dependent motor skill learning. *Neuron, 35*(1), 205–211.

Walker, M. P., Brakefield, T., Hobson, J. A., & Stickgold, R. (2003). Dissociable stages of human memory consolidation and reconsolidation. *Nature, 425*(6958), 616–620.

Part II
Commentaries

Chapter 5
Do You Still Hold to an Activation Only Theory of the REM Sleep-Dreaming Correlation?

John Antrobus

Allan's new paper, Dream Consciousness, represents the ideal function of theory, particularly in a new field such as the cognitive neuroscience of dreaming. A theory is not a belief. It is, rather, an abstract representation of what we know and might reasonably assume about the phenomena under consideration. A *good* theory is one that provokes enough new research for it to be revised or replaced – ideally within 5 years. This is precisely how Allan's AIM model trumps Freud's 1899 theory of dreaming. This new proposal that "REM sleep dreaming ... provides a structural and functional building block for waking consciousness..." is based on several revisions of the original AIM model that Allan originally published in *The Neuropsychology of Sleep and Dreaming* (1992). Revisions of the original model were made in response to continuous new research findings both within neurophysiology and human dreaming.

To appreciate Allan's contribution to the cognitive neuroscience of dreaming, one need only look at the huge volume of dream psychophysiology research that followed the REM-dreaming discovery in 1953. The field was dominated by a broad effort to locate isomorphic relations between psychophysiological markers and subjective dream experience. For example, perhaps a hundred experiments evaluated the hypothesis that the direction of eye movements in REM sleep corresponded to the direction of the dreamer's gaze. We assumed that the beta EEG of REM indicated and activated the cortex, and we knew that the active cortex was deafferented in REM sleep (Pompeiano 1967) but we lacked a unified neurophysiological model that might provide a comprehensive conceptual basis for sleep imagery and thought. In response to our compilation of the first 25 years of such dreaming research in *The Mind in Sleep* (Arkin et al. 1978), Allan complained in an otherwise kind review – but where is the theory? He rose to the challenge with his AIM model in which he has led the attempt to represent the neurophysiology and subjective experience of dreaming in a single coherent model.

J. Antrobus (✉)
Program in Cognitive Neuroscience, City College of New York, New York, NY, USA
e-mail: j4ss4j@aol.com

N. Tranquillo (ed.), *Dream Consciousness*, Vienna Circle Institute Library 3,
DOI 10.1007/978-3-319-07296-8_5,

Allan points to the tension between AIM theory and Continuity theory and suggests that some Continuity theorists "... seem (to) want to possess dream theory as a purely psychological process in order to protect it as intellectual territory ... (and are) ... either ignorant of, or downright hostile to physiology." But activation and sensory inhibition – a component of I – were the foundation of continuity theory even before AIM was proposed. AIM placed them in a unified subcortical neural foundation for activation and sensory inhibition from which it generated a host of new neurocognitive assumptions about dreaming. The challenge for AIM was to account for more than the simple cortical activation-deafferent model of dreaming.

Initial resistance came when extensive careful experimental tests failed to support a very plausible assumption of the I component. The original AIM assumed that the cortex needs an "input" to create a dream, inasmuch as the cortex in deafferent inputs is inhibited in REM sleep. In waking perception, pontine-geniculo-occipital (PGO) spikes are a component of eye movements (EMs). In REM sleep, PGO spikes also drive eye movements. A strong assumption of the original AIM model was that the REM sleep PGO spikes are interpreted by the activated cortex as eye movements responding to visual percepts. PGO spikes are the "input" (I) to the dream in that the process of "synthesizing" this EM information creates the visual imagery of the dream. But a series of well-designed experiments found no support for this perfectly reasonable hypothesis.

I assume, by contrast, that each active cortical region behaves as an attractor in that it attempts to "make sense" of whatever information or noise reaches it and, in 1991, I simulated this interpretive process in a neural network model. I assume that all active, interconnected (See, Braun et al. 1997) cortical regions continuously send noisy information to one another during REM sleep and they jointly attempt to "make sense" of whatever information or noise reaches them. Lacking accurate external information, their production is conventionally called hallucinatory. But this quality requires only activation (A) and the sensory deafferent component of I of AIM – not the M component.

And there are additional features that AIM attributes to modulation (M) that we find can be accounted for quite well by A and deafferentation. Both Allan and I quite reasonably tend to assume that every unique characteristic of REM dreams must have a unique neurophysiological basis. AIM attributes hallucinosis, bizarreness and other salient features of dreaming to M, on the assumption that they cannot be accounted for by A. But, again, hard empirical data fail to support these assumptions with reference to AIM's M component. The problem here is that we are all too confident that our own private experience provides us with a good representation of waking thought and imagery against which we can reliably construct REM sleep mentation. Despite the thousands of studies of sleep mentation, not one had ever studied the characteristics of the imagery and thought of waking subjects under the same conditions employed in laboratory-based sleep studies. When Reinsel, Wollman and I did so (Reinsel et al. 1992), we found, to our great surprise, that waking imagery and thought was significantly *more* vivid and *more* bizarre than in REM sleep! There were two differences. Waking fantasy was never hallucinatory

– a difference we attributed to the deafferent factor in REM sleep. Second, while waking thought was more bizarre in terms of abrupt changes in theme and location (see Rechtschaffen 1978), it was not more bizarre in terms of improbable identities, e.g. it was my brother but he was a girl. We attribute this latter class of bizarreness – rare even in REM sleep – to the frequent disconnect between cortical regions that interact closely in waking perception (Braun et al. 1997).

Like Allan, I agree that these differences as well as other quite plausible AIM assumptions can be resolved with further research. But neurocognitive dream research is very expensive. Following the Aserinsky and Kleitman discovery (1953), the National Institute of Mental Health (NIMH) poured a large part of its funding into dream research on the explicit assumption that it would pull psychoanalysts into science and possibly enhance clinical practice. With the loss of that generous research support it has been, and will continue to be, very difficult to continue the evaluation of the many creative assumptions of AIM as well as the additional issues suggested at the end of Allan's paper. For example, the role of the active amygdala in REM sleep offers a way to bring personal values back into models of dreaming. Although limited in its scope, Freud's great contribution was to emphasize that personal value – drive or motive – modifies cognitive process. The activation of amygdala-prefrontal cortex networks in REM sleep offer a potential basis for this value-affect-cognition interaction. Finally, the recent discovery of the default network (see, Mason et al. 2007) that identifies the cortical regions and pathways that participate in mind wandering and SIT (stimulus-independent thought) (Antrobus et al. 1970) in the waking state offers an articulated cortical model that may be similar to that of dreaming sleep – both within REM and EEG stage 2. I join Allan in anticipating how these new developments in cognitive neuroscience will advance the neuroscience of dreaming and consciousness.

References

Antrobus, J. S., Singer, J. L., Goldstein, S., & Fortgang, M. (1970). Mindwandering and cognitive structure. *Transactions of the New York Academy of Sciences, 32*(2), 242–252.

Arkin, A. M., Antrobus, J. S., & Ellman, S. J. (Eds.). (1978). *The mind in sleep: Psychology and psychophysiology*. Hillsdale: Lawrence Erlbaum Associates.

Aserinsky, E., & Kleitman, N. (1953). Regularly occurring periods of ocular motility and concomitant phenomena during sleep. *Science, 118*, 361–375.

Braun, A. R., et al. (1997). Regional cerebral blood flow throughout the sleep–wake cycle. *Brain, 120*, 1173–1197.

Hobson, J. A. (1992). A new model of brain-mind state: Activation level, input source, and mode processing (AIM). In J. Antrobus & M. Bertini (Eds.), *The neuropsychology of sleep and dreaming* (pp. 227–246). Hillsdale: Lawrence Erlbaum.

Mason, M. F., Norton, M. I., Van Horn, J. D., Wegner, D. M., Grafton, S. T., & Macrae, C. N. (2007). Wandering minds: The default network and stimulus-independent thought. *Science, 315*(5810), 393–395.

Pompeiano, O. (1967). The neurophysiological mechanisms of the postural and motor events during desynchronized sleep. *Research Publications – Association for Research in Nervous and Mental Disease, 45*, 351–423.

Rechtschaffen, A. (1978). The single-mindedness and isolation of dreams. *Sleep, 1*, 97–109.

Reinsel, R., Antrobus, J., & Wollman, M. (1992). Bizarreness in dreams and waking fantasy. In J. Antrobus & M. Bertini (Eds.), *The neuropsychology of sleep and dreaming* (pp. 157–184). Hillsdale: Lawrence Erlbaum Associates.

Chapter 6
What Is the Most Promising Avenue to New Understanding of the Sleep-Learning Process?

Mercedes Atienza and José L. Cantero

Neuroscientists have provided strong evidence in the last two decades that sleep contributes to building new memories. However, we are a long way from understanding the specific mechanisms responsible for this achievement. One general belief held by sleep researchers is that offline reactivation of experience-specific activity patterns during sleep makes newly acquired memories more resistant to interference, more resilient to the passage of time, and more integrated with pre-existing knowledge. The reactivation hypothesis states that similar neuronal patterns associated with encoding are spontaneously reproduced by the same neuronal assemblies during sleep when the original event is no longer present. Contrary to the general conviction, this neural replay is not specific to sleep, but it can also occur during waking, for instance just before free recall of a specific item (Gelbard-Sagiv et al. 2008). Unfortunately, neither in waking nor in sleep, causality between reactivation and memory consolidation has not been demonstrated to date. Recent advances with multivariate patterns classifiers trained to distinguish oscillatory neural responses associated with different stimuli will provide novel insights into this field (Fuentemilla et al. 2010). As these classifiers gain expertise, they will be able to detect learning-dependent neural patterns in the absence of sensory stimulation during sleep.

The fact that reactivation of learning-dependent neural patterns is feasible both in waking and sleep does not mean that sleep is not playing a specific role in memory formation via similar mechanisms. Thus, coupling between sleep oscillations, like hippocampal sharp-wave ripples (80–200 Hz), thalamo-cortical spindles and thalamo-cortical ultra slow waves (<1 Hz), might provide an optimized time window for reactivation and transfer of memories. Accordingly, convergent evidence relates changes in the above neural oscillations to memory enhancement following sleep. This framework suggests that sleep may not be the only milieu in

M. Atienza (✉) • J.L. Cantero
Laboratory of Functional Neuroscience, Spanish Network of Excellence for Research on Neurodegenerative Diseases (CIBERNED), University Pablo de Olavide, Seville, Spain
e-mail: matirui@upo.es; jlcanlor@upo.es

N. Tranquillo (ed.), *Dream Consciousness*, Vienna Circle Institute Library 3,
DOI 10.1007/978-3-319-07296-8_6,

which memory strengthening is possible, but it can facilitate and speed up the underlying processes. Again, the issue left to determine is whether cooperation between the neural systems subserving ripples, spindles and slow waves is causally related to memory consolidation processes. The application of transcranial slow oscillation stimulation in non REM sleep has already provided promising results relating this sleep-specific oscillatory activity to memory enhancement on the next day (Marshall et al. 2006), but more studies in this direction are clearly needed.

In the last few years, a new view of how sleep affects memory has gained popularity. It does not necessarily contradict the consolidation hypothesis, but complements it. In particular, it has been suggested that the specific role of sleep in memory function is to restore the encoding capabilities of synaptic connections. In support of this hypothesis, it has been found that a night of sleep deprivation (Yoo et al. 2007), or even decreases in the activity of slow waves (Van der Werf et al. 2009), is enough to interfere with the encoding of new memories. Likewise, evidence suggests that a daytime nap is able to improve memory for newly encoded information when compared with information encoded prior to sleep (Axmacher et al. 2008). However, sleep oscillatory patterns that have been shown to correlate with post-sleep enhanced memory were also found to predict post-sleep successful encoding. Therefore, novel analytic approaches are urgently needed to disentangle neural events associated with consolidation processes from those responsible for local homeostasis processes.

Novel insights into the role of sleep in memory consolidation are coming from neurodegenerative disorders associated with memory impairment. Along this line, we have found that persons at risk of developing Alzheimer's disease (AD) show a significant reduction of REM sleep (Hita-Yañez et al. 2012). Evidence suggests that neurons of the nucleus basalis of Meynert release the greatest level of acetylcholine to the neocortex during REM sleep, playing a major role in regulating cortical EEG activation characteristic of this cerebral state. Interestingly, these cholinergic neurons are also selectively vulnerable to neurodegeneration in AD and patients with mild cognitive impairment (Grothe et al. 2010). Although cholinergic dysfunctions could be partially responsible for REM deficits observed in AD patients, their impact on memory impairment remains elusive to date. Future research combining sleep, memory, and neurodegeneration may provide an alternative approach to better understand the complex relationship between sleep physiology and failed memories.

Collectively, all these research programs will shed light into the specific role of sleep in building new memories in our brains.

References

Axmacher, N., Haupt, S., Fernandez, G., Elger, C. E., & Fell, J. (2008). The role of sleep in declarative memory consolidation - direct evidence by intracraneal EEG. *Cerebral Cortex, 18*, 500–507.

Fuentemilla, L., Penny, W. D., Cashdollar, N., Bunzeck, N., & Düzel, E. (2010). Theta-coupled periodic replay in working memory. *Current Biology, 20*, 606–612.

Gelbard-Sagiv, H., Mukamel, R., Harel, M., Malach, R., & Fried, I. (2008). Internally generated reactivation of single neurons in human hippocampus during free recall. *Science, 322*, 96–101.

Grothe, M., Zaborszky, L., Atienza, M., Gil-Neciga, E., Rodriguez-Romero, R., Teipel, S. J., Amunts, K., Suarez-Gonzalez, A., & Cantero, J. L. (2010). Reduction of basal forebrain cholinergic system parallels cognitive impairment in patients at high risk of developing Alzheimer's disease. *Cerebral Cortex, 20*, 1685–1695.

Hita-Yañez, E., Atienza, M., Gil-Neciga, E., & Cantero, J. L. (2012). Disturbed sleep patterns in elders with mild cognitive impairment: The role of memory decline and ApoE ε4 genotype. *Current Alzheimer Research, 9*, 290–297.

Marshall, L., Helgadottir, H., Mölle, M., & Born, J. (2006). Boosting slow oscillations during sleep potentiates memory. *Nature, 444*, 610–613.

Van der Werf, Y. D., Altena, E., Schoonheim, M. M., Sanz-Arigita, E. J., Vis, J. C., De Rijke, W., et al. (2009). Sleep benefits subsequent hippocampal functioning. *Nature Neuroscience, 12*, 122–123.

Yoo, S. S., Hu, P. T., Gujar, N., Jolesz, F. A., & Walker, M. P. (2007). A deficit in the ability to form new human memories without sleep. *Nature Neuroscience, 10*, 385–392.

Chapter 7
Are You Convinced That Dreaming Is A Conscious State?

Susan Blackmore

No. Indeed, I am unhappy with the very notion of a "conscious state." States have to be states of something, so what is dreaming a state of? Here are some possible candidates:

> The whole organism. In this case, if someone is in REM sleep they are not conscious – at least they are unresponsive and so not conscious in that simple sense.

Hobson suggests that a "mental process" may be conscious. Assuming that we adopt his brain/mind equivalence, we now have to argue that some brain/mind processes are conscious while others are not. This leads us into a veritable minefield. The idea of conscious and unconscious processes is commonly accepted and the popular hunt for the neural correlates of consciousness (NCCs) is based on this distinction. Yet it entails the curious notion that some brain processes have this special additional property of being, or becoming, conscious while others do not – some require solutions to the "hard problem" of consciousness, while others do not. Dennett's classic *Consciousness Explained* (1991) can be seen as an attempt to demolish this idea. I have argued elsewhere (Blackmore 2009) that brain processes cannot be divided into conscious and unconscious ones. Indeed, I suggest it would entail a "magic difference" (Blackmore 2010). So I can make no sense of the idea that dreaming is a conscious state of a mind/brain process.

We might instead say that it is "me," the self or subject of experience, that is in a conscious state during dreams. Yet this is clearly untrue for ordinary dreams. After a typical dream, we wake up with memories of all the bizarre, interesting, scary, or peculiar things that were apparently going on but which stopped when "I" awoke. This is quite different from a lucid dream in which, as Hobson explains, the person knows that they are dreaming. The feeling of becoming lucid in a dream is often described as becoming more conscious, or "waking up," within a dream. In this sense, one might say that in lucid dreams the self is more conscious than in ordinary

S. Blackmore (✉)
School of Psychology, University of Plymouth, Plymouth, UK
e-mail: susan.blackmore@virgin.net

N. Tranquillo (ed.), *Dream Consciousness*, Vienna Circle Institute Library 3,
DOI 10.1007/978-3-319-07296-8_7,

dreams. But even this faces difficulties because we do not yet understand what a self is or could be. Without a better understanding of self-processes, self-models, illusions of self, or other ways of understanding what a self is, we cannot decide whether a self is in a conscious state or not.

I would like to add one further comment. Hobson claims that it is "profitable to regard dreaming not as unconscious but rather as an altered state of consciousness that is difficult to recall in waking" as though this were a new and bold idea. But it is not. Many elementary psychology textbooks define dreaming as "an altered state of consciousness occurring during sleep" and many consciousness researchers describe it as a conscious phenomenon: "Dreaming is a subjective phenomenon of consciousness" (Revonsuo and Tarkko 2002, p. 4); "Dreams are conscious because they create the appearance of a world . . . Dreams are subjective states in that there is a phenomenal self" (Metzinger 2009, p. 135). "Dreams are a form of consciousness, though of course quite different from full waking states" (Searle 1997, p. 5).

According to Hobson, "Our dreams are not mysterious phenomena, they are conscious events" (Hobson 1999, p. 209). But what does this mean? The notion of "conscious events" is itself a great mystery – perhaps the greatest mystery facing science – and Hobson's theory, despite making valuable contributions in other respects, does nothing to solve it.

7.1 What Impact Does the Apparent Loss of Volition in Dreams Have on the Free Will Debate?

This loss of volition is fascinating, not least because volition reappears in lucid dreams. It seems that when a dreamer becomes aware that they are dreaming, not only does their sense of self become much more like their waking self, but they can often take control of the dream events and of their own actions in the dream. As Hobson has explained, this change appears to be associated with greater activity in some parts of the frontal lobes which is what we would expect given the role of these brain areas in waking volition.

I think this tells us something about the connection between self and volition. When a good enough self-model is constructed, actions are attributed to that self. So the illusion of free will arises along with the illusion of self.

To answer the question – this interesting fact ought to have an impact on the free will debate but I doubt that it will. The illusion of conscious will (Wegner 2002) is so powerful that for hundreds of years people have preferred to follow their intuition that "of course I have free will" than accept what appears to be obvious from science that our will is a function of brain activity and is not free at all. However, I hope that in combination with other such findings we may eventually find that the common intuition of free will may be left behind, as we long ago left behind such notions as caloric fluid, the luminiferous ether and the life force.

References

Blackmore, S. (2009). *Ten zen questions*. Oxford: OneWorld.

Blackmore, S. (2010). *Consciousness: An introduction* (2nd ed.). London: Hodder Education.

Dennett, D. C. (1991). *Consciousness explained*. London: Little, Brown & Co.

Hobson, J. A. (1999). *Dreaming as delirium: How the brain goes out of its mind*. Cambridge, MA: MIT Press.

Metzinger, T. (2009). *The ego tunnel: The science of the mind and the myth of the self*. New York: Basic Books.

Revonsuo, A., & Tarkko, K. (2002). Binding in dreams: The bizarreness of dream images and the unity of consciousness. *Journal of Consciousness Studies, 9*(7), 3–24.

Searle, J. (1997). *The mystery of consciousness*. New York: New York Review of Books.

Wegner, D. M. (2002). *The illusion of conscious will*. Cambridge, MA: MIT Press.

Chapter 8
Your PET Studies Demonstrate Major Differences Between REM Sleep and Waking. How Do You Conceive of These Differences in Relation to Theories of Waking and Dreaming Consciousness?

Allen R. Braun

Allan Hobson presents a clear and compelling introduction to a new theory of protoconsciousness – a distillation and extension of his earlier, seminal body of work. He takes a tripartite Jamesian approach, considering dream consciousness from psychological, physiological and philosophical perspectives.

Fundamental are the sections that summarize the psychological and physiological data. These make a strong case that the psychological features of dreams can be demystified and fully understood in physiological terms, putting another nail in the coffin of Freud's notion that dreams guard sleep against the intrusion of unconscious wishes, by means of censorship and symbolic disguise. (This particular notion seems to keep escaping from its coffin, but perhaps its days are numbered).

Refreshingly though, Hobson is ecumenical in spirit; he leaves open the possibility that unconscious material may emerge into dream consciousness and that psychotherapists may make use of this – but need not resort to classical psychoanalytic interpretation in order to do so.

Nor is he a reductionist but, as a "dual aspect monist," Hobson grounds his approach in objective scientific inquiry and its requisite third person validation, while making a compelling case for introspection and first person experience as legitimate material for such an inquiry ("responsible introspectivism").

He places protoconsciousness in a philosophical perspective in the final section of his monograph and goes on to trace a number of its theoretical implications.

The most compelling aspect of the theory, and what appears to be its central tenet, is the claim that rather than reflecting unconscious mental activity, dreaming represents an altered state of consciousness *per se*. The corollary – that as such it may provide a means of understanding the evolution of consciousness itself – is a bold one.

A.R. Braun (✉)
National Institute on Deafness and Other Communication Disorders, National Institutes of Health, Bethesda, MD, USA
e-mail: brauna@nidcd.nih.gov

N. Tranquillo (ed.), *Dream Consciousness*, Vienna Circle Institute Library 3,
DOI 10.1007/978-3-319-07296-8_8,

The theory is new however and as such is still incomplete; a few elements are at present underspecified and may need to be addressed.

This seems to me to be most evident in the discussion on clinical phenomena – particularly the idea that the hallucinations, delusions, disorientation, single mindedness, and amnesia that characterize dreams are also the major symptoms of mental illness.

This statement may be a bit broad. Hobson has pointed out in the past that the visual, attentional and mnemonic features of dreaming are remarkably consistent with the clinical features of delirium – but they are less consistent with those of other common illnesses, such as schizophrenia or bipolar disorder.

Even if it were true that the primary features of dreams are typical of psychiatric illnesses, the theory should probably specify the mechanisms by which these features emerge in the mentally ill. If the "hardware for psychosis is built into the brain," exactly how is it constrained or regulated during wake in normal individuals? How are "endogenous visual images suppressed during wake to prevent hallucinations?" And what's gone wrong in the patients?

Can this be explained in terms of the AIM model? Do external stimuli keep internal inputs at bay during wake? What is the role of neuromodulation? Do higher ambient levels of norepinephrine (NE) and serotonin (5-HT) prevent the emergence of psychotic behavior and hallucinosis? If so, why is hallucinosis not a behavioral consequence of monoaminergic blockade?

Do other neurotransmitters play a role? What about orexin? Would the transition from wake to REM in narcoleptics – when waking consciousness lurches abruptly into protoconsciousness (if I understand this correctly) – provide some clues about this pathophysiological trigger? What do these patients report at the moment of transition?

Another set of questions relates to Hobson's notion that a principal objective of the mind-brain is to construct an *a priori* model of the world – and that this may explain some of the perceptual and narrative features of dreams. That is, these features are a result of perceptual filling in and narrative construction, primed by *a priori* experience.

This is an interesting notion, but I wonder what the theory has to say about the content of (what we might assume to be) dreaming *in utero* or during early infancy, periods rife with REM sleep.

What happens then, when perceptual systems are inchoate, when internal signals cannot be formally matched with signals that have been coming from the outside world? What should dream consciousness look like without expectant codes or stimulus identifier circuits to organize perceptual filling in or narrative synthesis?

A few additional questions related to dream emotionality: Exactly how can protoconsciousness theory account for the fact that the lion's share of dream emotions appear to be negative – e.g. fear and anger – in humans as well as, apparently, Jouvet's cats? Why such selectivity? There is a suggestion that this may be due to evolutionary pressures. Exactly how?

And, since dreams are clearly social, why are sadness, loneliness, shame, guilt – affects that Hobson identifies as social – largely absent? Hobson suggests that these are linked to analytic, heteromodal brain systems (which are off-line during REM), but surely they are mediated by the same emotional brain systems that are activated during REM. Can the theory account for this dissociation?

These are largely minor points. Protoconsciousness is potentially important and its primary thesis – that REM sleep dreaming is a conscious rather than an unconscious state – is an important paradigm. More than simply making a strong case for "taking sleep and dream science seriously," Hobson lays the groundwork for a research agenda that would progressively define the mechanisms and roles of this enigmatic third state of consciousness. Protoconsciousness is inchoate however; some of the remaining loose threads need to be interwoven and the theory itself may require more rigorous specification.

I confess that I cannot add much to this. I will simply comment on what functional neuroimaging studies may have to offer to the scientific investigation of consciousness – and suggest that, after the dramatic breakthroughs in studies of sleep over the past decade and a half, neuroimaging still has much to offer. Specifically, I think results that have emerged in studies of waking subjects can serve as important points of reference in ongoing and future studies of sleep and dreaming.

In particular, I am referring to imaging-based investigations of the so-called default mode network (DMN) – a set of higher order heteromodal cortical areas that include the medial prefrontal cortex, posterior cingulate/retrosplenial cortices, inferior parietal lobules and portions of the temporal lobe.

This network has been associated with a wide variety of proposed roles, incorporating what appear to be disparate behavioral and cognitive domains. The DMN was first described in association with internally directed processes – introspection, rumination, stimulus-independent thinking and the like – all of which are typically studied in the "resting-state."

But the DMN plays a role in externally-directed cognitive processes as well – including social cognition and "theory or mind," spatial navigation, and storytelling.

This may sound like a pastiche of unrelated functions, but I think there is a unifying principle that suggests that the DMN plays a larger role in consciousness and intentionality – and that this aid our understanding of self-awareness during both waking and dreaming.

To digress for a moment: I'd like to make use of John Searle's definition of consciousness – which is compatible with Hobson's preferred dual-aspect monism. Both suggest that consciousness is a natural, biological phenomenon that is instantiated in the brain, and that it can be legitimately investigated using both first and third person perspectives.

Searle believes that consciousness is a primary, necessary feature of any mental state – i.e. mind cannot exist without it. A corollary – and what makes Searle's conception unique – is an obligatory relationship between consciousness and intentionality.

What is intentionality? It is the proper occupation of consciousness and conscious awareness. It arises when our conscious mental states – our thoughts, sensations, beliefs, desires – are about something. It is central to the first person point of view.

If intentionality describes the conscious relationship between the self and the world, it is important to note that this can be the external or internal world. That is, our thoughts, beliefs and desires can, for Searle, be about the environment – the world out there – or can be directed at an internal representation of the world. The contents of consciousness can be real or imaginary, external or internal, past, present, or future.

So what does this have to do with the default mode network? Well, I think all of the proposed roles outlined above, all of the activities in which the DMN appears to be engaged – from introspection to social understanding to storytelling – can be united under the rubric of intentionality as Searle defines it.

The overarching role of the DMN may be in mediating – perhaps via its secondary connections distributed throughout the brain – the relationship between the self and the objects of conscious awareness.

It may be responsible for what Hobson defines as one of the principal emergent features of subjectivity: the "perception of the world with a self at the center of that world." Once again – it can be the social world "out there," or an internal representation of the world, which can be accessed during resting, wakefulness, or perhaps during dreaming.

What makes all of this interesting, and potentially relevant, is the following:

While the initial presumption was that the DMN was always active, functionally integrated in all instances – including unconsciousness – evidence to the contrary has begun to accumulate. It appears instead that DMN integrity varies significantly with the level of consciousness.

For example, the groundbreaking work emerging from the laboratory of Steven Laureys has shown that DMN activity is absent in brain death, coma and vegetative state, but emerges to some degree in patients who are minimally conscious. But this is incomplete, with only partial activation of anterior portions of the network in these patients.

And crucially, using standard BOLD fMRI connectivity methods, Silvina Horovitz and colleagues have demonstrated that the anterior and posterior nodes of the DMN are robustly uncoupled – in much the same fashion as Laurey's minimally conscious patients – during slow wave sleep. [Indeed, the work of Renate Wehrle, Michael Czisch and co-workers has shown that DMN integrity has already begun to dwindle during extended periods of sleep deprivation].

Together, these findings suggest an intriguing possibility: that integrity of the DMN may reflect, and perhaps even constitute, 'conscious awareness' *per se*. The mechanism? Laureys and Brent Vogt have speculated that the DMN's central posterior node, the posterior cingulate and contiguous retrosplenial cortices, may integrate activity in subcortical systems that regulate arousal and cognitive systems distributed throughout the cerebral cortex.

Perhaps the DMN in this way integrates the primary and secondary features of consciousness and, through its widespread interactions throughout the forebrain, organizes the contents of consciousness. These connectivity patterns may define the features of any conscious, intentional state.

So here are some predictions:

If the DMN is uncoupled during the deepest stages of slow wave sleep, we expect that it will be re-coupled during REM.

But since REM and waking consciousness differ so dramatically, it is likely that the regions that comprise the DMN will be connected to other cortical and subcortical areas in a fundamentally different way, and that a thoroughly unique connectivity pattern will define this third state of consciousness.

In terms of intentionality, it is of course not the outer world with which the newly "awakened" self makes contact during REM, but a world characterized by the internally generated, hallucinatory, confabulatory features of dreams.

The functional implications of this – of what Hobson identifies as the spontaneity, unpredictability, chaos, randomness of dreaming – will be discussed below.

Of course this is pure speculation (and I may be running the risk of proposing the DMN as a pineal for the twenty-first century). But we may make even more detailed predictions – specifically that relationships between the DMN's principal anterior node, the medial prefrontal cortex, and other prefrontal regions will characterize dreams associated with REM sleep – and here we have actual data to guide us.

In this case, it will be helpful to revisit REM imaging data that have already been published – albeit in large part overlooked.

Although what is generally commented upon is deactivation of the dorsolateral prefrontal cortex (DLPFC), REM is in fact associated with concomitant activation of the medial prefrontal cortex (MPFC), to levels that may actually exceed those seen during wake. This has been reported by our lab as well as by those of Pierre Maquet (using PET) and Andreas Ioannides (using MEG).

So it is not simply that the prefrontal cortex is "offline" during REM. Instead a functional dissociation between medial and lateral portions of the prefrontal cortex may be one of the hallmarks of REM and of the dreams that accompany it.

This is interesting in light of additional data from our lab. We have evidence that creative artistic behavior, in particular free improvisation by jazz artists, is associated with increased activity in the MPFC and decreased activity in the DLPFC – precisely the pattern seen during REM sleep.

We interpreted this pattern in the following way: that activation of MPFC indexed intentionality in action, expression of the self in a free, unstructured state made possible by suspension of critical self monitoring that is normally mediated by the DLPFC.

Freedom from these restraints, it might be argued, permits novel associations to be made by the artist, who is able to capitalize on the "spontaneity, unpredictability, chaos, and randomness" that characterize dreaming.

If creativity and dreams share a common neural substrate, it seems that the Surrealists' intuitions were correct (even if they were ultimately misdirected by psychoanalytic thought).

We reported this pattern in musicians, but have recently seen the same reciprocal relationship between MPFC and DLPFC during spontaneous storytelling. The parallel with Hobson's ideas about dream plot selection is obvious, and is in general consistent with the spirit of his theory. The "unpredictability that goes hand in hand with chaos" characteristic of REM sleep dreaming may be harnessed by artists who are, in common parlance, "in the zone." (And, if Hobson is correct, this may provide some insights into the possible relationships between dreaming, creativity and madness).

Most importantly, this suggests a straightforward mechanism by which patterns characteristic of REM can be accessed during wake – and vice versa.

A simple reorganization of cortical connectivity, perhaps involving the DMN, may underlie the transmutation between states of consciousness. This may be evident in the balance between activity in the medial and dorsal portions of the prefrontal cortex – where increased DLPFC activity during REM sleep may enable lucid dreaming, and decreased DLPFC activity during wake may give rein to creative improvisation.

This is entirely consistent with Hobson's notion "that dream consciousness ... depends upon the enhancement of some, and the weakening of other, aspects of waking consciousness."

In this view, perhaps both REM and wake may be separate and equally complex states of consciousness, built upon different configurations of cerebral connections. But they are states that can shift and slide into one another, consistent with Hobson's notion that waking and dreaming consciousness are reciprocal and interactive: sometimes positively, enabling inspiration and insight; sometimes pathologically, unleashing waking delirium.

In conclusion, although he may be premature in saying that the mind-brain problem is essentially solved, I admire Hobson's confidence, and am convinced by his insistence that a resolution will come, in part, by using the concepts and methods of sleep and dream science.

I would venture to suggest that the continued application of neuroimaging methods will take us well beyond the initial breakthroughs of the past decade or so ago and will be of value in approaching the resolution that Hobson envisions.

Chapter 9
How Does Your Formulation of Lesion-Induced States of Diminished Consciousness Fit with AIM? Do You Suppose That Brain Stem Damage Affects Activation (A) and Modulation (M)?

Vanessa Charland-Verville and Steven Laureys

9.1 The AIM Model, Coma, and Related States

Allan Hobson's AIM model (Hobson 1998) is built according to three main dimensions. The first component, *Activation*, describes the brain's activation processes and is closely linked to the level of consciousness. According to the model, the brain is highly active in wakefulness and REM sleep but will show much less activity during NREM sleep. The second component, *Input/output gateway*, controls the inhibition of external stimuli. When slowly falling asleep, the gateway shuts down and inhibits the external stimuli; the brain is no longer involved in processing external perceptions. Then the brain starts its oniric phase and the focus switches to internal inputs. Finally, the third dimension, *Modulation*, refers to the different ways of cognitive processing (executive functions), judgment, volition and memory. According to the model, those cognitive processes are lacking in REM sleep (i.e., the brain cannot keep a record of its conscious experience during dreaming as opposed to waking state) because of the changes between the aminergic system (norepinephrine and serotonin; dominant in waking but ineffective in REM sleep) and the cholinergic system (acetylcholine; unfettered in REM sleep). These three dimensions maintain a dynamic and reciprocal interaction over the sleep-wake cycle's variations (wakefulness, NREM, and REM sleep) and each of them can be expressed with lower or higher intensities depending on the level of consciousness.

V. Charland-Verville (✉) • S. Laureys
Coma Science Group, Cyclotron Research Centre and Neurology Department, University of Liège, Liège, Belgium
e-mail: vanessa.charland-verville@ulg.ac.be

N. Tranquillo (ed.), *Dream Consciousness*, Vienna Circle Institute Library 3,
DOI 10.1007/978-3-319-07296-8_9,

9.2 Disorders of Consciousness (DOC) and the AIM Model

For neurologists, consciousness can be defined according to two main components: arousal (the level of consciousness) and awareness (the content of consciousness) (Zeman 2001). The arousal component involves the activity of a complex network comprising the cholinergic reticulothalamic projections; glutaminergic thalamocortical projections; and the reticulocortical projections (a network of dopaminergic, noradrenergic, serotonergic, and cholinergic projections), whereas the awareness component involves the connectivity of a widespread thalamo-cortical fronto-parietal network (Laureys et al. 2000). In normal physiological states, arousal and awareness are linearly correlated that is, the more awake we are the more we are conscious (Laureys 2005). Awareness can in turn be subdivided into external (e.g., stimulus-dependent thoughts; sensation and perception of environmental external stimuli) and internal awareness (e.g., stimulus-independent thoughts; inner speech, mental imagery, daydreaming, and mind wandering) (Vanhaudenhuyse et al. 2011). Recent advances in medicine and reanimation have considerably increased the number of patients surviving anoxic injuries or severe brain trauma. Of those survivors, some of them will not recuperate fully and will stay temporarily or chronically in a state of altered consciousness (Laureys and Boly 2008).

9.2.1 *Disorders of Consciousness*

The majority of patients will recover from their coma within the first days after the insult, while others will take more time and go through different stages before fully or partially recovering awareness (e.g., minimally conscious state, vegetative state/unresponsive wakefulness syndrome) or will permanently lose all brain functions (i.e., brain death).

Brain death is defined as an irreversible state of unconsciousness with the complete and irreversible loss of brainstem and brain (Guidelines for the Determination of Death 1981). Indeed, since the neurological criteria of brain death were first described, no patient in apneic coma who was properly declared brain (or brainstem) dead has ever regained consciousness (Pallis 1983; Wijdicks 2001).

Coma is defined as a condition of "unarousable unresponsiveness" and can result either from bihemispheric diffuse cortical white matter damage or brain stem lesions, affecting the subcortical reticular arousing systems. Clinically, comatose patients show no spontaneous eye opening and cannot be awakened by external stimulation; arousal and awareness are absent and they show no sleep-wake cycles (Posner et al. 2007).

Vegetative state, or now preferably referred to as *Unresponsive wakefulness syndrome* (VS/UWS; Laureys et al. 2010), represents a unique dissociated state of consciousness since it is characterized by normal arousal levels with a preserved sleep-wake cycle (without REM sleep; Cologan et al. 2010), yet without awareness (Laureys 2005; The Multi-Society Task Force on PVS 1994). The VS/UWS results from widespread cortical or white matter and/or bilateral thalamic lesions. The brain stem autonomic functions are sufficient for survival with adequate medical assistance (Demertzi et al. 2009).

Minimally conscious state (MCS) patients show inconsistent but reliable signs of external awareness that can be reproduced and/or sustained long enough to be differentiated from reflexive behaviors (Giacino et al. 2002). By definition, MCS patients cannot communicate their thoughts and wishes. This state has recently been subdivided into two independent subcategories based on clinical behaviors and neuroimaging data; MCS *plus* (MCS+) patients show response to simple commands, while MCS *minus* (MCS-) patients demonstrate non-reflexive signs of consciousness that are not related to language processing (e.g., visual pursuit or localization to pain) (Bruno et al. 2011). Compared to MCS+, MCS- patients may suffer from a significant general decrease in brain metabolism in the left hemisphere and particularly in regions that are functionally linked to speech comprehension and production, in motor and pre-motor areas, and in sensory-motor areas. Differential diagnosis for MCS would therefore be mainly due to the functional recovery (or not) of speech processing areas (Bruno et al. 2012).

Finally, *locked-in syndrome* (LIS) patients are awake and aware of themselves and their environment but are unable to produce movements or communicate, except for eye coded-communication. After a period of coma, they will regain normal arousal and awareness levels but will remain paralyzed and voiceless due to a ventral pontine lesion producing supranuclear motor de-efferentation preventing any voluntary movements of all four limbs and the last cranial nerves without interfering with consciousness (Plum and Posner 1983). At the bedside, LIS patients can superficially resemble VS/UWS patients (Bruno et al. 2009; Laureys 2005; Smart et al. 2008).

9.3 Disorders of Consciousness in the AIM Space

9.3.1 Activation and Measuring the Level of Consciousness in DOC

The first dimension of the AIM model is defined by brain active processes, which are linked to the amount of brain energy consumption. For example, activation is present while awake or dreaming but drops substantially during NREM sleep. Clinically, the brain's metabolic activity is an important biomarker in assessing the level of consciousness in brain-damaged patients (Laureys et al. 1999a, b; Phillips et al. 2010). This validated measure of energy consumption can be determined with neuroimaging techniques like [18F]-fluorodeoxyglucose positron emission tomography (FDG-PET) imaging. FDG-PET studies in brain death show the absence of any metabolic activity except for the portion of the skin surrounding the skull (i.e., the "empty skull sign"; Laureys et al. 2004). In coma and VS/UWS, the global cerebral metabolism decrease is about 50–70 % while MCS patients demonstrate preserved metabolic activity in the consciousness network encompassing fronto-parietal regions and midline structures (Bruno et al. 2012; Laureys et al. 2004). Finally, LIS patients show a comparable to normal overall brain metabolic activity.

Interestingly, resting neuroimaging FDG-PET studies have shown that the recovery of global metabolic activity is not a prerequisite for regaining consciousness. Rather, it has been shown that a fronto-parietal network encompassing midline (i.e., anterior cingulate/mesiofrontal and posterior cingulate/precuneus) and lateral (i.e., prefrontal and posterior parietal) associative cortices is consistently hypometabolic in VS/UWS patients and regains metabolic activity when the patient recovers signs of awareness (Laureys et al. 1999a, b; Lull et al. 2010). Data obtained in sleep (for a review e.g., Maquet 2010) and general anesthesia (for a review e.g., Boveroux et al. 2008) also corroborate these findings.

9.3.2 Input/Output Gateway in Disorders of Consciousness

According to the AIM model, the information will be processed differentially depending on its source. External data processing is possible when the sensory input and motor output gates are open (i.e., in waking consciousness). On the contrary, for internal information to be processed (i.e., dream content) input/output gates are closed and internal stimulus generation is enhanced. As previously mentioned, awareness can be divided into two networks or sources: awareness of the environment (external) and of the self (internal). External awareness refers to the conscious perception of one's environment through the sensory modalities (e.g., visual, auditory, somesthetic, or olfactory perception). Internal awareness refers to mental processes that do not require the mediation of external stimuli or sensory input (e.g., mind wandering, daydreaming, inner speech, or mental imagery). Interestingly, it has been shown that healthy controls' behavioral reports of internal awareness were linked to the activity of midline anterior cingulate/mesiofrontal areas as well as posterior cingulate/precuneal cortices, while subjective ratings for external awareness correlated with the activity of lateral fronto-parieto-temporal regions (Vanhaudenhuyse et al. 2011). Moreover, these two independently distributed awareness networks showed an anticorrelated pattern between each other, being of functional relevance to conscious cognition.

In DOC, FDG-PET imaging in VS/UWS shows a widespread metabolic dysfunction in internal and external networks and thalami. MCS tends to show less impairment in the external/lateral network than in the internal/medial network. This observation is in line with the fact that these patients show evidence of external/ sensory awareness, known to depend upon the functional integrity of the extrinsic/

lateral fronto-parietal network at the bedside during clinical behavioral examinations (Boly et al. 2007a; Sergent and Dehaene 2004; Vanhaudenhuyse et al. 2011). The predominance of internal/midline network impairment in MCS patients could reflect an impairment in self-awareness, which is very difficult to quantify at the bedside. More specifically, MCS+ patients show a recovery of brain metabolism in right and left temporo-parietal regions while MCS- patients mainly recover metabolic activity in the right hemisphere. Finally, in LIS patients, both the extrinsic and intrinsic network activity is preserved with impairment only at the cerebellum level (Thibaut et al. 2012).

In addition, increasing attention has been paid to the study of spontaneous brain activity and its significance for cognition and behavior. The "default mode network," associated with the internal input/awareness network, shows more activity at rest than during attention-demanding tasks. In other words, the activity of these regions is higher when having internal thoughts than when interacting with the external environment. This "default mode network" activity has been associated with the level of consciousness in brain damaged non-communicative patients using fMRI. In fact, the network's functional connectivity correlates with the level of consciousness ranging from healthy controls to MCS, VS/UWS, and coma patients showing that internal awareness input decreases with the level of consciousness (Vanhaudenhuyse et al. 2010).

9.3.3 Modulation

According to Hobson's AIM model, modifications in cognition are dependent on the ratio changes in the concentrations of neuromodulators like acetylcholine, serotonin, and norepinephrine. Neurochemical changes in DOC are very challenging to assess and human data are still very difficult to gather. On the other hand, some evidence shows that consciousness can be improved following specific neuropharmacological treatment. For instance, after a 5-week single case therapeutic trial of Amantadine (a dopaminergic agonist), a drug-related metabolic increase in fronto-parietal regions, as well as significant improvements in cognitive functions, have been observed in a chronic post-traumatic MCS patient after the administration of the drug (Schnakers et al. 2008a). Furthermore, at the group level, a recent controlled trial with placebo in post-traumatic DOC patients (VS/UWS and MCS) showed that the administration of Amantadine also significantly enhanced the rate of functional recovery (e.g., functional behaviors such as consistent responses to commands, intelligible speech, reliable yes/no communication, and functional-object use) over a 4-week period of treatment, as compared with a placebo trial. Another pharmacological agent, Zolpidem (a non-benzodiazepine GABA agonist used to treat occasional insomnia), has been shown to have paradoxical positive neurological effects on VS/UWS and MCS patients in the restoration of brain functions (Brefel-Courbon et al. 2007; Shames and Ring 2008; Whyte and Myers 2009).

As well as causing modifications at the neurotransmitter level, anoxic and traumatic diffuse axonal injuries to the connections of the arousal system with the thalamo-cortical fronto-parietal network may impair cognition and consciousness to varying degrees. In order to assess cognitive functions and, in particular, meaningful goal-directed behaviors, in moderately to severely brain-injured non-communicative patients, recent functional neuroimaging studies have provided the possibility not only to detect signs of consciousness, but also to observe higher cognitive functioning via command following specific brain activations. In fact, studies using active paradigms with fMRI provided evidence for awareness and higher cognitive processing in patients behaviorally diagnosed with MCS and UWS/VS and as they presented with volitional brain activity (Bardin et al. 2011; Boly et al. 2007b; Monti et al. 2010; Owen et al. 2006). These results stress the possible dissociation between responsiveness and preserved consciousness. Comparable paradigms are also used with the electroencephalogram (Cruse et al. 2011; Goldfine et al. 2011; Schnakers et al. 2008b) or electromyogram (Bekinschtein et al. 2008).

Hobson's three-dimensional state space AIM model describes brain-related processes as a way of visualizing the consciousness states (i.e., mainly waking, NREM, and REM sleep states). With the model parameters encompassing (1) *Activation*, (2) *Input/output gateway*, and (3) *Modulation*, it is possible to follow conscious states as a continuous succession of points in a particular space. The aim of the present contribution was to explore the interaction of these brain parameters together with recent findings in pathological cases of altered consciousness. Brain lesions leading to DOC will affect each of the model's parameters. However, DOC patients still represent a very challenging field in medicine and neuroscience and the underlying mechanisms of the AIM model can be challenging to identify. Furthermore, our understanding of the sleep-wake cycles in DOC patients is also very incomplete (Cologan et al. 2010). In conclusion, we believe that by combining theoretical and explanatory correlates of consciousness, science will bring insights into the proper understanding of conscious cognition and its recovery after acquired brain injury.

References

Bardin, J. C., Fins, J. J., Katz, D. I., Hersh, J., Heier, L. A., Tabelow, K., Dyke, J. P., Ballon, D. J., Schiff, N. D., & Voss, H. U. (2011). Dissociations between behavioural and functional magnetic resonance imaging-based evaluations of cognitive function after brain injury. *Brain, 134*(Pt 3), 769–782.

Bekinschtein, T. A., Coleman, M. R., Niklison, J., 3rd, Pickard, J. D., & Manes, F. F. (2008). Can electromyography objectively detect voluntary movement in disorders of consciousness? *Journal of Neurology, Neurosurgery, and Psychiatry, 79*(7), 826–828.

Boly, M., Balteau, E., Schnakers, C., Degueldre, C., Moonen, G., Luxen, A., Phillips, C., Peigneux, P., Maquet, P., & Laureys, S. (2007a). Baseline brain activity fluctuations predict somatosensory perception in humans. *Proceedings of the National Academy of Sciences of the United States of America, 104*(29), 12187–12192.

Boly, M., Coleman, M. R., Davis, M. H., Hampshire, A., Bor, D., Moonen, G., Maquet, P. A., Pickard, J. D., Laureys, S., & Owen, A. M. (2007b). When thoughts become action: An fMRI paradigm to study volitional brain activity in non-communicative brain injured patients. *NeuroImage, 36*(3), 979–992.

Boveroux, P., et al. (2008). Brain function in physiologically, pharmacologically, and pathologically altered states of consciousness. *International Anesthesiology Clinics, 46*(3), 131–146.

Brefel-Courbon, C., Payoux, P., Ory, F., Sommet, A., Slaoui, T., Raboyeau, G., Lemesle, B., Puel, M., Montastruc, J. L., Demonet, J. F., & Cardebat, D. (2007). Clinical and imaging evidence of zolpidem effect in hypoxic encephalopathy. *Annals of Neurology, 62*(1), 102–105.

Bruno, M. A., Schnakers, C., Damas, F., Pellas, F., Lutte, I., Bernheim, J., Majerus, S., Moonen, G., Goldman, S., & Laureys, S. (2009). Locked-in syndrome in children: Report of five cases and review of the literature. *Pediatric Neurology, 41*(4), 237–246.

Bruno, M. A., Vanhaudenhuyse, A., Thibaut, A., Moonen, G., & Laureys, S. (2011). From unresponsive wakefulness to minimally conscious PLUS and functional locked-in syndromes: Recent advances in our understanding of disorders of consciousness. *Journal of Neurology, 258*(7), 1373–1384.

Bruno, M. A., Majerus, S., Boly, M., Vanhaudenhuyse, A., Schnakers, C., Gosseries, O., Boveroux, P., Kirsch, M., Demertzi, A., Bernard, C., Hustinx, R., Moonen, G., & Laureys, S. (2012). Functional neuroanatomy underlying the clinical subcategorization of minimally conscious state patients. *Journal of Neurology, 259*(6), 1087–1098.

Cologan, V., Schabus, M., Ledoux, D., Moonen, G., Maquet, P., & Laureys, S. (2010). Sleep in disorders of consciousness. *Sleep Medicine Reviews, 14*(2), 97–105.

Cruse, D., Chennu, S., Chatelle, C., Bekinschtein, T. A., Fernandez-Espejo, D., Pickard, J. D., Laureys, S., & Owen, A. M. (2011). Bedside detection of awareness in the vegetative state: A cohort study. *Lancet, 378*(9809), 2088–2094.

Demertzi, A., Laureys, S., & Boly, M. (2009). *Coma, persistent vegetative states, and diminished consciousness*. Oxford: Elsevier.

Giacino, J. T., Ashwal, S., Childs, N., Cranford, R., Jennett, B., Katz, D. I., Kelly, J. P., Rosenberg, J. H., Whyte, J., Zafonte, R. D., & Zasler, N. D. (2002). The minimally conscious state: Definition and diagnostic criteria. *Neurology, 58*(3), 349–353.

Goldfine, A. M., Victor, J. D., Conte, M. M., Bardin, J. C., & Schiff, N. D. (2011). Determination of awareness in patients with severe brain injury using EEG power spectral analysis. *Clinical Neurophysiology, 122*(11), 2157–2168.

Guidelines for the Determination of Death. (1981). *JAMA, 246*(19), 2184–2186.

Hobson, J. A. (1998). *Consciousness*. New York: Scientific American Library.

Laureys, S. (2005). The neural correlate of (un)awareness: Lessons from the vegetative state. *Trends in Cognitive Sciences, 9*(12), 556–559.

Laureys, S., & Boly, M. (2008). The changing spectrum of coma. *Nature Clinical Practice Neurology, 4*(10), 544–546.

Laureys, S., Goldman, S., Phillips, C., Van Bogaert, P., Aerts, J., Luxen, A., Franck, G., & Maquet, P. (1999a). Impaired effective cortical connectivity in vegetative state: Preliminary investigation using PET. *NeuroImage, 9*(4), 377–382.

Laureys, S., Lemaire, C., Maquet, P., Phillips, C., & Franck, G. (1999b). Cerebral metabolism during vegetative state and after recovery to consciousness. *Journal of Neurology, Neurosurgery & Psychiatry, 67*(1), 121.

Laureys, S., Faymonville, M. E., Luxen, A., Lamy, M., Franck, G., & Maquet, P. (2000). Restoration of thalamocortical connectivity after recovery from persistent vegetative state. *Lancet, 355*(9217), 1790–1791.

Laureys, S., Owen, A. M., & Schiff, N. D. (2004). Brain function in coma, vegetative state, and related disorders. *Lancet Neurology, 3*(9), 537–546.

Laureys, S., et al. (2010). Unresponsive wakefulness syndrome: A new name for the vegetative state or apallic syndrome. *BMC Medicine, 8*(1), 68.

Lull, N., Noe, E., Lull, J. J., Garcia-Panach, J., Chirivella, J., Ferri, J., Lopez-Aznar, D., Sopena, P., & Robles, M. (2010). Voxel-based statistical analysis of thalamic glucose metabolism in traumatic brain injury: Relationship with consciousness and cognition. *Brain Injury, 24*(9), 1098–1107.

Maquet, P. (2010). Understanding non rapid eye movement sleep through neuroimaging. *World Journal of Biological Psychiatry, 11*(Suppl 1), 9–15.

Monti, M. M., Vanhaudenhuyse, A., Coleman, M. R., Boly, M., Pickard, J. D., Tshibanda, L., Owen, A. M., & Laureys, S. (2010). Willful modulation of brain activity in disorders of consciousness. *The New England Journal of Medicine, 362*(7), 579–589.

Owen, A. M., Coleman, M. R., Boly, M., Davis, M. H., Laureys, S., & Pickard, J. D. (2006). Detecting awareness in the vegetative state. *Science, 313*(5792), 1402.

Pallis, C. (1983). ABC of brain stem death. The declaration of death. *British Medical Journal, 286* (6358), 39.

Phillips, C. L., Bruno, M.-A., Maquet, P., Boly, M., Noirhomme, Q., Schnakers, C., Vanhaudenhuyse, A., Bonjean, M., Hustinx, R., Moonen, G., Luxen, A., & Laureys, S. (2010). "Relevance vector machine" consciousness classifier applied to cerebral metabolism of vegetative and locked-in patients. *NeuroImage, 56*(2), 797–808.

Plum, F., & Posner, J. B. (1983). *The diagnosis of stupor and coma*. Philadelphia: FA Davis.

Posner, J. B., Saper, C. B., & Plum, F. (2007). *Diagnosis of stupor and coma*. New York: Oxford University Press.

Schnakers, C., Hustinx, R., Vandewalle, G., Majerus, S., Moonen, G., Boly, M., Vanhaudenhuyse, A., & Laureys, S. (2008a). Measuring the effect of amantadine in chronic anoxic minimally conscious state. *Journal of Neurology, Neurosurgery & Psychiatry, 79*(2), 225–227.

Schnakers, C., Perrin, F., Schabus, M., Majerus, S., Ledoux, D., Damas, P., Boly, M., Vanhaudenhuyse, A., Bruno, M. A., Moonen, G., & Laureys, S. (2008b). Voluntary brain processing in disorders of consciousness. *Neurology, 71*(20), 1614–1620.

Sergent, C., & Dehaene, S. (2004). Neural processes underlying conscious perception: Experimental findings and a global neuronal workspace framework. *Journal of Physiology, Paris, 98* (4–6), 374–384.

Shames, J. L., & Ring, H. (2008). Transient reversal of anoxic brain injury-related minimally conscious state after zolpidem administration: A case report. *Archives of Physical Medicine and Rehabilitation, 89*(2), 386–388.

Smart, C. M., Giacino, J. T., Cullen, T., Moreno, D. R., Hirsch, J., Schiff, N. D., & Gizzi, M. (2008). A case of locked-in syndrome complicated by central deafness. *Nature Clinical Practice Neurology, 4*(8), 448–453.

The Multi-Society Task Force on PVS. (1994). *The New England Journal of Medicine, 330*(22), 1572–1579.

Thibaut, A., Bruno, M. A., Chatelle, C., Gosseries, O., Vanhaudenhuyse, A., Demertzi, A., Schnakers, C., Thonnard, M., Charland-Verville, V., Bernard, C., Bahri, M., Phillips, C., Boly, M., Hustinx, R., & Laureys, S. (2012). Metabolic activity in external and internal awareness networks in severely brain-damaged patients. *Journal of Rehabilitation Medicine, 44*(6), 487–494.

Vanhaudenhuyse, A., Noirhomme, Q., Tshibanda, L. J.-F., Bruno, M.-A., Boveroux, P., Schnakers, C., Soddu, A., Perlbarg, V., Ledoux, D., Brichant, J.-F., Moonen, G., Maquet, P., Greicius, M. D., Laureys, S., & Boly, M. (2010). Default network connectivity reflects the level of consciousness in non-communicative brain-damaged patients. *Brain, 133*(1), 161–171.

Vanhaudenhuyse, A., Demertzi, A., Schabus, M., Noirhomme, Q., Bredart, S., Boly, M., Phillips, C., Soddu, A., Luxen, A., Moonen, G., & Laureys, S. (2011). Two distinct neuronal networks mediate the awareness of environment and of self. *Journal of Cognitive Neuroscience, 23*(3), 570–578.

Whyte, J., & Myers, R. (2009). Incidence of clinically significant responses to zolpidem among patients with disorders of consciousness: A preliminary placebo controlled trial. *American Journal of Physical Medicine and Rehabilitation, 88*(5), 410–418.

Wijdicks, E. F. (2001). The diagnosis of brain death. *The New England Journal of Medicine, 344*(16), 1215–1221.

Zeman, A. (2001). Consciousness. *Brain, 124*(Pt 7), 1263–1289.

Chapter 10
Does Your "Feeling of What Happens" Definition of Consciousness Extend to Dreaming? If So, How Do You Conceptualize Internally Generated FWHs?

Antonio Damasio

Allan Hobson's lectures provide two distinct rewards. First, a review of important ideas and facts from his research on dreams. Second, the lectures present an account of consciousness that roots the phenomenon in the dream process. I like Hobson's account very much and agree with most, though not all, of Hobson's choices.

Let me begin my comments with the ideas that I am entirely comfortable with. Hobson unequivocally considers REM sleep dreams as states of consciousness. I concur wholeheartedly. I also agree with Hobson that relegating dreams into an unconscious sector of processing is unjustified and unhelpful, notwithstanding the fact that dream contents can be influenced by non-conscious events. My agreement is complete because Hobson is careful to define REM sleep dreams as states of *altered* consciousness, a qualification that is often left out when dreams are simply described as conscious without further comment. The state of consciousness that prevails in REM dreams is anything but standard. Standard consciousness requires wakefulness. When wakefulness is suspended and the brain is functionally isolated—it can neither be acted upon by external stimulations, nor act on the organism's surround—the mind's contents become indeed strange and logic is relaxed, though not necessarily abolished. In other words, dream consciousness is not normal consciousness, but it is consciousness nonetheless. I often insist on precisely this same point. Regrettably, this has been interpreted as my saying that dreams are not conscious mind states. They evidently are, complete with an ingredient I regard as indispensable for the conscious label to apply: a self process. The self process of dream states can be especially robust in lucid dreaming but it is still present in other dreams as well. Absent a self process, how would experience be possible? Experiences require an *experiencer*, based on the property of subjectivity that selves confer upon the mind.

A. Damasio (✉)
Brain and Creativity Institute, University of Southern California, Los Angeles, CA, USA
e-mail: damasio@usc.edu

N. Tranquillo (ed.), *Dream Consciousness*, Vienna Circle Institute Library 3, 111
DOI 10.1007/978-3-319-07296-8_10, © Springer International Publishing Switzerland 2014

I also agree with Hobson's defense of the emotional congruence of dreams. In most (if not all) dreams, the signal of the emotion is appropriate to the dreamed events. On the other hand, I do not agree that one's regular waking experience is emotionless. We are fortunately spared large emotional upheavals for long stretches of daily living, but I believe there is a continuously fluctuating state of background emotion and feeling. I regard the complete absence of that emotional background as pathological.

Another suggestion of Hobson's that I especially like is that REM sleep dream states resemble the hallucinatory states of awake individuals in whom consciousness is usually regarded as normal. This has important implications for research on the psychoses.

Let me now turn to Hobson's use of REM sleep dreams as a basis for an account of the origins of consciousness. Here, I agree with the notion that dream consciousness is likely to be a less evolved stage of consciousness than standard wakeful consciousness. However, I resist Hobson's idea that dreams represent the bottom level of the consciousness process, that they are the protoconsciousness, as he calls it. If I understand his idea correctly, the earliest conscious mind states would be dream-like and would mark the very evolutionary beginning of conscious mind processing. My resistance to this comes from the fact that I see the beginning of consciousness as occurring far earlier in biological history and in physiological level. I see protoconsciousness as coinciding with simple events that I describe as primordial feelings. Primordial feelings, in my account, regard the representation of ongoing states of the organism. Such feelings manifest existence, in a powerful way, but they lack the rich imagetic contents of dreams. They are not connected to the representation of external objects and events.

As for the anatomical and physiological level associated with the content of dreams, it is certainly that of the cerebral cortex working in partnership with subcortical structures. This is not so for the primordial feelings that ground my notion of protoconsciousness. They originate at the level of the humble brainstem and hypothalamic nuclei.

In conclusion, I agree with Hobson that dream states probably represent forerunners of standard conscious mind states. I also agree that they probably stand for a stage in the evolution of conscious processing. But I prefer to think that dreams are not the most primitive of conscious states. I place protoconsciousness at a level far more modest than that of dreams.

Chapter 11
What Is Dreaming for, If Anything?

Daniel C. Dennett

One of the charges leveled against adaptationism (most famously by Gould and Lewontin 1979, in their attack on "just-so stories") is that we adaptationists jump to our "panglossian" hypotheses without due consideration, let alone testing, of alternative explanations of the phenomena under discussion. The charge is not without merit, but I have always viewed it as pointing to a foible of adaptationists, not a fundamental flaw in adaptationism. So as a devout adaptationist, one of my duties is to police the brethren for just such lapses. Allan Hobson provides a nice instance, since he is cautious and circumspect in his consideration of the claim that dreaming has an important function to perform, and then he *still* overlooks or underplays alternatives. First, I will note what Hobson has to say about the (apparent) function of dreaming. Then, I will list just a few alternative hypotheses, to illustrate my point, and ask whether Hobson has given us any evidence that rules them out. According to Hobson:

> It makes good sense to prepare the brain for subsequent waking and the physiology of REM sleep manifests abundant signs of an activation state favoring the massive and parallel connectivity essential to the binding of the multiple cognitive processes that are required to explain the extraordinary unity of waking consciousness. For this reason, I refer to REM sleep dreaming as a proto-conscious state.

It does indeed "make good sense" but that is not enough. We need to see that disrupting/removing REM sleep has a deleterious effect, and even showing that would not be a slam dunk for function, since however we suppress REM, it may be a side-effect of suppression, rather than the lack of REM, that causes whatever decrement in cognitive competence we observe. What we really need is an account of just what REM sleep does to "integrate" the brain, and that could only be speculative at this time. It is certainly tempting to suppose, as Hobson does, that

D.C. Dennett (✉)
Center for Cognitive Studies, Tufts University, Medford, MA, USA
e-mail: Daniel.Dennett@tufts.edu

N. Tranquillo (ed.), *Dream Consciousness*, Vienna Circle Institute Library 3,
DOI 10.1007/978-3-319-07296-8_11,

> REM sleep constitutes an elaborate program for sensorimotor integration, emotion evocation, and scenario construction. It may be hard to prove but it is hard to believe that these features are not in the service of waking consciousness. The fact that this state recurs every night of our lives underscores that assertion.

It underscores the assertion, but leaves plenty of room for alternatives. For instance:

(a) Dreaming in infancy has a function, but dreaming in adulthood is simply a vestige of a phenomenon that did indeed play a crucial organizing role in the early days of every infant's (or fledgling's) neural development. What function? REM in the womb is (roughly) what Hobson says REM is in the adult: the spontaneous, endogenous activity that shapes functional structures in the absence of current sensory input, a sort of practice perception off-line that helps prepare the neophyte brain for the real world. But once the brain has a regular diet of sensory input to structure it, the off-line activation becomes obsolete but is harmless and so not worth suppressing. It recedes in duration to only 1.5 h a day on average, a fraction of its duration in infancy. On this hypothesis, adult dreaming is rather like adult thumbsucking, a functionless habit that was itself a byproduct of an infantile imperative of considerable functional importance.
(b) Same as A, except that REM is not in fact so functional during infancy and gestation. It is simply that the brain gets "turned on" in development before it has anything to work with, so it idles vigorously, not doing any harm, but also not doing much if any pre-birth organization, having so little to go on, aside from quite a lot of auditory, tactile and vestibular input.
(c) REM could also be, as Hobson suggests, a byproduct of some neural process involved in thermoregulation. The fact that it has evolved twice is an important datum, but we should be cautious about leaping beyond "byproduct" to "function." Again, we'd want to know *how* REM supported thermoregulation before taking this giant step.

We also need to be cautious about back-extrapolating our connotations of adult dreaming onto the infant and fetal phenomena—a sort of intraspecific anthropocentric bias! Just because we human adults report dreams that are brimming with fascinating content is not a particularly good reason to infer that infants (or birds) are "enjoying the show" during their REM periods. But on what grounds, other than bloody-minded philosophical skepticism (more on that below), would lead us to reserve judgment about infant *experience* during their REM periods? For one thing, the existence of a feedback loop between (adult) human communication and REM, nicely established by Hobson. As he notes, it is quite possible to train yourself to have lucid dreams, or to wake up and write down dream narratives when they occur. This loop is of course entirely absent in dogs and birds and other languageless REMsters, including pre-linguistic human infants. How could this capacity to be trained have arisen in prehistory and how could it have shaped the phenomenon of dreaming in turn? Here is one way in which dreaming (with content, with narratives, as a proto-conscious experience) could have evolved:

> In phase one, there is REM sleep, as described in A or B, and it has some interesting side effects in our ancestors once they begin to develop language: youngsters occasionally blurt out words or longer fragments of speech on awakening (or they talk in their sleep) and their parents respond with interest (curiosity and concern) and this creates the feedback loop that encourages further such blurting, which elaborates and crystallizes over time into dream reportage. There might be a long period (phase two) when only a few people ever exhibit these symptoms even though they all have REM sleep. During this phase there would not be a word for "dream" or a concept of dreaming, and it would be a bit of a stretch to say that dreaming was, already, a phenomenon in this early human world, not just because they didn't have a word for it, but because the feedback loop was not yet well enough elaborated to generate anything worth calling dreams. In a similar spirit, we can envisage a time when our ancestors had not yet managed to invent/discover the phenomena of *homesickness* or *fame* or *war* or *pornography*.

From this perspective, which could be fleshed out in a variety of ways, there is a significant amount of content shaping by the demand characteristics of parental (and then self-) curiosity, not unlike the *folie à deux* elaborations of what used to be called Multiple Personality Disorder but is now officially known as Dissociative Identity Disorder (Humphrey and Dennett 1989). The reward (to put it crudely) of attention can, as Hobson shows, reach down beneath the surface of verbal expression and begin to modulate the processes that feed verbal expression, opening the floodgates to more and more *tellable* dreams. So we would have something of a continuum between REM episodes with no organized content at all, REM episodes that create vague content-laden hunches, emotional squalls of one flavor or another, for instance, and on to fully ripened dream experiences. The capacity to "recollect" dreams would thus be an artifact of cultural conditioning rather than a functional enhancement of an underlying functional process. Dreams wouldn't have to be *for* anything—except for telling, and that activity could be a culturally infectious habit, harmless and even entertaining, and hence hard to extinguish.

It would be true of any of these hypotheses that children in learning to speak also are learning to dream, a process in which the opportunity/task of recalling their dreams when awake plays an encouraging role. Then cultural evolution or contagion could take over, and we're on the yellow brick road to seers and shamans, prophets and interpreters, Jung and Freud.

This brings us close to Hobson's challenge to me to clarify my position on Norman Malcolm's notorious hypothesis about dreams. A re-reading of my 1975 paper "Are Dreams Experiences?" (Dennett 1978) shows that I never defended Norman Malcolm's "hypothesis that dream reports reflect mental activity associated with the awakening process rather than antecedent REM or NREM sleep physiology." (Hobson, personal communication) Rather, I used Malcolm's outrageous but ingenious proposal to expose some of the otherwise tacit and underappreciated assumptions that must be taken on board by dream researchers—assumptions that Hobson and others have indeed committed to, without, perhaps, recognizing that they subtly undermine some "common sense" ideas about dreams and indeed all conscious experience.

Chief among these, no doubt, is the quite standard image of dreams as experiences that occur in strict narrative sequence, quite like a movie running in the head while the sleeper has REM. The fact that on awakening dreamers *report* such putative episodes is undeniable, but not as conclusive as some would suppose. There are, for instance, temporal anomalies (from the point of view of this standard image), and these appear paradoxical *until* you abandon the standard image and acknowledge that the temporal properties of neural events and the temporal properties *represented* by neural events are entirely independent, in principle, no matter how closely yoked they are *in waking life*—aside from such interesting and telltale cases as *color phi*, *metacontrast* and the *cutaneous rabbit* (Dennett 1991). The basic physics of living in the world, with events occurring not just in sequence, but continuously, without major hiatuses or leaps in time and space, disciplines our everyday waking perception by imposing deadlines for behavioral control. This is a major feature of conscious experience, and dream processes, occurring under a more relaxed regime, need not obey these constraints. These processes still must reside in the evolved, trained machinery of the perceptual systems of the brain, and here Hobson's model of a partly chaotic, noise-driven (instead of information-driven) elaboration of normal perceptual analysis processes is a nice confirmation and detailed elaboration of my sketchier version, in the prelude, "How are Hallucinations Possible?" to *Consciousness Explained* (Dennett 1991, esp. pp. 10–16.)

Hobson stresses, and I concur, that the line between conscious and proto-conscious or non-conscious is not to be drawn in a principled way. That *in itself* is such a departure from everyday thinking about consciousness and experience that Malcolm stands almost vindicated, if we interpret him as defending the thesis that dreams are not experiences *in the ordinary sense of the term*. (Remember, he was an Ordinary Language philosopher writing in 1959.) I do not want to defend Malcolm's antique view (Malcolm 1959), since the last half century of research, especially by Hobson, has deepened our understanding tremendously, and Malcolm was blinkered by his verificationism. As I noted in 1975, Malcolm hugely underestimated the power of models and theories (Dennett 1975). My brand of verificationism is, I think, more supple, and more attuned to the demands of actual science: we must be cautious when the urge arises to export conclusions *couched in ordinary language* from scientific investigations. Hobson *is* cautious; he recognizes that he has a considerable diplomatic task confronting him, clearing away the rubbish of several obsolete creeds that still impede communication: the "functional vs. organic" dichotomy, the anti-neurophysiological bias of many psychologists and therapists, and the granddaddy of them all: Freud. I applaud his polemics against all these prejudices, but I do wonder sometimes whether he has outlived his adversaries. Are there really influential Freudians still out there dragging their heels? Since in the field I know from the inside, philosophy of mind, I see Hobson sometimes pounding on an open door, I have some grounds for this suspicion. Brain-mind unification has pretty well gone to fixation among philosophers, with dualism now the renegade position, the outlier—which means, I fear, that a few young philosophers are tempted to endorse it just to be naughty and notorious (such is our field, alas).

In general, Hobson's philosophical views on the nature of the mind and consciousness would be welcomed as reassuringly mainstream by my fellow philosophers. They ought to be discomfited, however, by a few signs in Hobson's essay of what I think of as overgenerous applications of the principle of charity. In several instances Hobson takes himself to be addressing philosophers' concerns when in fact his comments, sound in themselves, don't reach any targets in philosophy land, only because philosophers are not talking about anything as interesting as the topic he is discussing. They are talking about a trivial artifactual puzzle of their own devising. For instance, Hobson's "Hard Problem" is not the philosophers' Hard Problem, thank goodness. The philosophers' Hard Problem is couched in such a way that no amount of brilliant theory about how the brain accomplishes one cognitive competence or another even *approaches* it. (Roughly, any problem you can solve is, by definition, one of the easy problems! The philosophers' Hard Problem systematically eludes all of cognitive neuroscience. Enough said.) Another instance that makes me cringe is his remark about "Dennett's zombie country." No, philosophers' zombies are *not* like "automatons" as Hobson imagines—those are the zombies of folklore. Philosophers' zombies are a much sillier idea: folks who are excellent company, lively, creative, lovable, as knowledgeable and sensitive as you could want—but *there's nobody home*. (This is well nigh impossible to imagine, but many philosophers have persuaded themselves that they can imagine it, and that their ability to do this feat of imagination is a Major Problem for the sciences of the mind.) When I inveigh against the Zombic Hunch (Dennett 2005), I am not criticizing the idea of automaton-like elements in our brains; I actually support that idea in detail (cf. Christof Koch's proposals on this score). Hobson doesn't in fact have anything to say about philosophers' zombies, I'm happy to report, since the topic is too embarrassing to merit consideration.

References

Dennett, D. C. (1975). Are dreams experiences? *Philosophical Review, 73*, 151–171. reprinted in Dennett, 1978.

Dennett, D. C. (1978). *Brainstorms: Philosophical essays on mind and psychology*. Montgomery: Bradford Books.

Dennett, D. C. (1991). *Consciousness explained*. Boston: Little, Brown.

Dennett, D. C. (2005). *Sweet dreams: Philosophical obstacles to a science of consciousness*. Cambridge, MA: MIT Press.

Gould, S. J., & Lewontin, R. (1979). The spandrels of San Marco and the panglossian paradigm: A critique of the adaptationist programme. *Proceedings of the Royal Society, B205*, 581–598.

Humphrey, N., & Dennett, D. C. (1989). Speaking for our selves: An assessment of multiple personality disorder. *Raritan: A Quarterly Review, IX*, 68–98. reprinted in Dennett, 1998.

Malcolm, N. (1959). *Dreaming*. London: Routledge & Kegan Paul.

Chapter 12
Your Interest in Sleep and Memory Posits a Semantic Learning Function to NREM Sleep. What Are Your Views of Associative Memory Enhancement by REM?

Susanne Diekelmann and Jan Born

Allan Hobson proposes a unique and highly inspiring model in which he composes a protoconsciousness during REM sleep by merging the physiological processes of REM sleep with introspections and formal analyses of dreaming. His "dream consciousness theory" serves a bouquet of new ideas on how dream-consciousness and wake-consciousness might be interrelated, and also provides a flavour of the (adaptive) function REM-related protoconsciousness provides for waking consciousness, an issue of central interest to sleep research in the post-Darwin area. As we understand, Hobson suggests that the random and chaotic nature of brain activation during REM sleep allows for creating novel and creative associations in memory.

Thus, tracing the role of sleep for memory with the instruments of experimental psychology and neurophysiology, we are now confronted by the question: is there an enhancement of associative memory in REM sleep? Previously, we proposed that it is NREM sleep, specifically slow wave sleep (SWS), rather than REM sleep, that takes the lead in the formation of semantic long-term memories (Diekelmann and Born 2010). However, we also suggested REM sleep might serve distinct but complementary functions for memory consolidation in a sequential process, although admittedly so far empirical evidence for this sequential hypothesis is shaky. It is known that SWS supports processes of system consolidation, i.e., the reactivation and reorganization of previously acquired memories. These memories are thereby gradually integrated into the network of long-term memories residing primarily in neocortical networks. During this process of reactivation and reorganization, invariant features shared by different episodes in memory are extracted to form a generalized context-independent semantic knowledge representation. Once such newly generalized memory traces are formed and adapted to the knowledge

S. Diekelmann (✉) • J. Born
Department of Neuroendocrinology, University of Lübeck, Lübeck, Germany

Department of Medical Psychology and Behavioral Neurobiology, University of Tübingen, Tübingen, Germany
e-mail: susanne.diekelmann@uni-tuebingen.de; born@kfg.uni-luebeck.de

N. Tranquillo (ed.), *Dream Consciousness*, Vienna Circle Institute Library 3,
DOI 10.1007/978-3-319-07296-8_12,

network, ensuing REM sleep might then function to support processes of synaptic consolidation, i.e., synaptic plastic changes to create enduring connections for the long-term. Obviously, in this concept, semantic memory representations are formed to the largest part in SWS rather than REM sleep. SWS is a state by far less clearly linked to any subjective experiences of dreams than REM sleep. Also, there is probably no protoconsciousness during SWS. Nevertheless, our concept seems well in line with Hobson's general theorizing assuming basically a profound control of wake consciousness by physiological processes going on in altered states of consciousness during sleep. Of note also, the memories formed during this dark phase of consciousness in SWS are well and explicitly accessible to waking consciousness.

As mentioned, the REM sleep part of our sequential SWS-REM sleep model of memory consolidation is clearly in need of both firmer empirical foundation and further theoretical sharpening – regarding the latter we can take advantage of Hobson's perspective. In fact, his approach adds a further captivating aspect to the memory function of sleep which he links to REM sleep specifically and which goes beyond classical memory function. The reason humans produce REM sleep dreams is not to provide a more or less correct but sparse copy of the past, but to synthesize, out of randomness, an "*a priori* model of the world," to create new associations to build new hypotheses that may eventually serve to control the waking future. Leaving aside the question whether Immanuel Kant would have appreciated this kind of *a priori* world where the brain out of prefrontal executive control forgets about his main *a priori* categories 'time', 'space', and 'causality', the important point in Hobson's theory is that the brain activation during REM sleep adds stochastic input to the brain network. One could easily expect that such random activation is detrimental to freshly encoded and previously reactivated memory traces. However, quite contrary to this notion, the noisy REM sleep activation might even benefit the consolidation of recent memories – by stimulating synaptic consolidation preferentially at synapses tagged during prior SWS – and even more so the forming of new associations with these memories. Although counterintuitive at first sight, the phenomenon that random noise can actually improve the quality of a signal is well known in physics and neurocomputing and named "stochastic resonance" (Gammaitoni et al. 1998; McDonnell and Abbott 2009).

Stochastic resonance refers to the fact that adding noise to a system can improve the probability of detecting a sub-threshold signal because the noise can increase the probability of the signal crossing the threshold. The closer an input is to the threshold, the more likely it will reach the threshold when noise is added. Especially new and still weak neuronal connections would profit from such noise whereby the overall signal-to-noise ratio can be maximized. Thus, by adding unspecific noise to the brain neuronal network, REM sleep might improve pre-existing weak signals including the representations that were reorganized during prior SWS (by allowing these signals at the synaptic level) to reach a certain threshold for synaptic consolidation processes to be initiated. Unfortunately, so far no compelling experimental evidence exists confirming that random activation in REM sleep, by facilitating such synaptic consolidation, indeed stabilizes memory associations in the long run to a greater extent than waking brain activity.

Whatever the reason is for this lack of evidence, the more important function of noisy REM sleep brain activation might be to generate novel signals or connections that have rarely or never been activated during the preceding wake phase or SWS periods but become activated during REM sleep merely by chance. The preferential activation of weakly over strongly associated items in semantic networks in REM-like brain states has in fact been experimentally demonstrated (Stickgold et al. 1999). Such strengthening of very weak connections might give rise to completely new and creative associations, and as Hobson's model suggests, random REM sleep activation might eventually set possible conceivable "memory states" that then "await" real sensory input. Such connections would virtually wait for future activation in a "prepared" state in which they are easily and rapidly excitable once corresponding external or internal stimuli are received. In this way, weakly associated stimuli activating such prepared representations might trigger creative ideas. This preparation of prospective probable associations might be especially important in the developing brain.

Some recent studies have suggested that insight can be fostered by SWS (Yordanova et al. 2008, 2010), whereas others found REM sleep that followed NREM sleep was implicated in such creative processes (Cai et al. 2009). Both lines of evidence can be reconciled when considering the proposed functions of SWS and REM sleep and the distinct aspects of convergent and divergent thought which are involved in the process of creative problem solving. SWS is expected to primarily benefit convergent features in this process that profit from a reactivation and reorganization of memory representations acquired prior to sleep, such as the extraction of invariant and repetitive features in stimulus patterns, the integration of new with similar older memories, etc. Chaotic brain activation however, during ensuing REM sleep, might specifically support divergent processes by allowing the formation of new connections and combinations among associative elements that were previously not or only very weakly associated, concurrently activating representations more or less by chance and thereby creating novel and original associations. Indeed, chaotic brain activation does not only hallmark REM sleep but also divergent (but not convergent) thinking (Molle et al. 1996, 1999). Thus, it is not associative memory but the associative play with memories that is benefited by REM sleep.

References

Cai, D. J., Mednick, S. A., Harrison, E. M., Kanady, J. C., & Mednick, S. C. (2009). REM, not incubation, improves creativity by priming associative networks. *Proceedings of the National Academy of Sciences of the United States of America, 106*, 10130–10134.

Diekelmann, S., & Born, J. (2010). The memory function of sleep. *Nature Reviews Neuroscience, 11*, 114–126.

Gammaitoni, L., Hänggi, P., Jung, P., & Marchesoni, F. (1998). Stochastic resonance. *Reviews of Modern Physics, 70*, 223–287.
McDonnell, M. D., & Abbott, D. (2009). What is stochastic resonance? Definitions, misconceptions, debates, and its relevance to biology. *PLoS Computational Biology, 5*, e1000348.
Molle, M., Marshall, L., Lutzenberger, W., Pietrowsky, R., Fehm, H. L., & Born, J. (1996). Enhanced dynamic complexity in the human EEG during creative thinking. *Neuroscience Letters, 208*, 61–64.
Molle, M., Marshall, L., Wolf, B., Fehm, H. L., & Born, J. (1999). EEG complexity and performance measures of creative thinking. *Psychophysiology, 36*, 95–104.
Stickgold, R., Scott, L., Rittenhouse, C., & Hobson, J. A. (1999). Sleep-induced changes in associative memory. *Journal of Cognitive Neuroscience, 11*, 182–193.
Yordanova, J., Kolev, V., Verleger, R., Bataghva, Z., Born, J., & Wagner, U. (2008). Shifting from implicit to explicit knowledge: different roles of early- and late-night sleep. *Learning and Memory, 15*, 508–515.
Yordanova, J., Kolev, V., Wagner, U., & Verleger, R. (2010). Differential associations of early- and late-night sleep with functional brain states promoting insight to abstract task regularity. *PLoS ONE, 5*, e9442.

Chapter 13
Your Imaging Group Has Coined the Term 'Dream Imaging.' Please Summarize the Concept in Relation to Dream Theory

Martin Dresler, Victor Spoormaker, Renate Wehrle, and Michael Czisch

The rise of modern imaging techniques has endowed neuroscience with a new quality to enquire into cognition and consciousness. In the 1990s, positron emission tomography (PET) was utilised for the first time to address sleep-specific questions and led to intriguing insights into the neural correlates of sleep. While the brain experiences widespread deactivations during slow wave sleep, REM sleep is again associated with increased cerebral blood flow in the thalamus, visual areas and limbic regions as well as attenuated metabolism in the dorsolateral prefrontal cortex (DLPFC), parietal cortex and the precuneus (Maquet et al. 1996, 1997; Braun et al. 1998). Such a pattern of activation has been related to the phenomenology of REM sleep dreaming. Activation of visual association areas is in line with vivid dream imagery, amygdala activation with dream emotionality, and DLPFC deactivation with the lack of insight, reflexivity and volition commonly experienced in dreams (Hobson and Pace-Schott 2002; Nir and Tononi 2009; Desseilles et al. 2011).

In contrast to PET, functional magnetic resonance imaging (fMRI) repeatedly allows non-invasive measurements of neural activity changes with high spatial resolution. However, fMRI faces several technical problems when applied to the study of sleep. While early fMRI sleep studies used behavioural measures to determine if the subject had fallen asleep (e.g. Hong et al. 2009), the development of MR-compatible EEG recording systems allowed us to obtain polysomnography in strong magnetic fields. Improvements in EEG postprocessing techniques have made it possible to substantiate that fMRI data were recorded during verified, unambiguous sleep (Czisch and Wehrle 2010). Consistent with PET findings, fMRI-measured cerebral activity in general decreases throughout NREM sleep as compared to wakefulness (Kaufmann et al. 2006). On the background of this globally decreased cerebral activity, dynamic increases in specific regional activity can be observed. In particular, neural correlates of sleep EEG microprocesses like

M. Dresler (✉) • V. Spoormaker • R. Wehrle • M. Czisch
Max Planck Institute of Psychiatry, Max Planck Society, Munich, Germany
e-mail: dresler@mpipsykl.mpg.de

N. Tranquillo (ed.), *Dream Consciousness*, Vienna Circle Institute Library 3,
DOI 10.1007/978-3-319-07296-8_13,

slow-waves (Dang-Vu et al. 2008), K-complexes (Czisch et al. 2009; Jahnke et al. 2012; Caporro et al. 2012), and sleep spindles (Schabus et al. 2007; Andrade et al. 2011; Caporro et al. 2012) have been analysed using fMRI. Fast sleep spindles have been shown to be paralleled by increased neural activity e.g. in the thalamus, cingulate and prefrontal cortex as well as pre- and postcentral gyrus, while slow spindles show a similar, but much weaker activity pattern (Schabus et al. 2007; Andrade et al. 2011). It was further shown that during spindle activity, processing of external acoustic stimuli is suppressed (Schabus et al. 2012), while the functional coupling between the hippocampus and the neocortex is increased (Andrade et al. 2011). These findings may reflect a spindle specific brain state allowing for increased information transfer between brain regions related to memory processes, with external interference being minimized.

External acoustic (Czisch et al. 2002, 2004; Dang-Vu et al. 2008) or visual (Born et al. 2002) perturbations during sleep have been used to characterize arousal reactions, but also sleep stabilizing mechanisms. When acoustic stimuli were continuously applied (for about half a minute) in NREM sleep stage 2, the brain's response showed a deactivation not only in the primarily targeted auditory areas, but nearly throughout the entire cortex (Czisch et al. 2002, 2004). Visual stimuli, on the other hand, led to a deactivation close to the visual cortex in the cuneus (Born et al. 2002). In the case of acoustic stimulation, deactivation was paralleled by an increase of slow EEG frequencies suggesting sleep deepening. Such deactivation may therefore be interpreted as a sleep protective mechanism reducing the processing of perturbing but non-alarming stimuli. Using the presentation of individual tones during light sleep, deactivation of motor cortical regions and the amygdala was confirmed (Czisch et al. 2009). In the latter analysis, it was further shown that whenever a tone presentation evoked a K-complex, the brain was transiently in a state allowing for stimulus processing and engaging cerebral regions typically activated during novelty processing.

Classical neurophysiological studies suggest that during REM sleep, the brain functions as a closed loop system in which activation is triggered in pontine regions while sensory input is gated and motor output is suppressed (Hobson and Pace-Schott 2002). Cortical synthesis of internally generated sensorimotor perceptions was proposed to explain partially coherent narrative dream mentation (Hobson and McCarley 1977). Until recently, the triggering ponto-geniculo-occipital (PGO) waves of this model were traceable in the animal model only. With fMRI, by correlating the appearance of eye movements and BOLD signal fluctuations, it was possible to directly visualise PGO-related activity in human sleep (Wehrle et al. 2005; Miyauchi et al. 2009).

The classical definition of REM sleep relies on the intermittent presence of REMs, which signify transient phasic activations embedded in tonic periods with sustained high-frequency EEG activity and suppression of muscle tone (Rechtschaffen and Kales 1968). Combined fMRI and polysomnography revealed differences in sensory processing and in thalamocortical activity patterns during tonic and phasic REM sleep periods in humans (Wehrle et al. 2007). While acoustic stimulation elicits residual activations of the auditory cortex during tonic REM sleep, periods containing bursts of phasic REM activity are characterized by a lack of reactivity to sensory stimuli. This difference is mirrored by different arousal thresholds during tonic and phasic REM sleep (Ermis et al. 2010). In addition, a thalamocortical network including limbic and parahippocampal areas is specifically active during phasic REM periods (Wehrle et al. 2007). Thus, REM sleep may have to be subdivided into tonic REM sleep with residual external stimulus processing, and phasic REM sleep with the brain acting as a functionally isolated and closed intrinsic loop. This neural difference might explain why external stimuli are sometimes incorporated into the dream narrative while at other times they are ignored or lead to awakening.

Another important but still under-recognised distinction besides tonic and phasic REM sleep is that between lucid and non-lucid sleep. During lucid dreaming, the sleeping subject becomes aware of his dreaming state, has full access to memory, and is able to volitionally control dreamed actions (LaBerge et al. 1981) while all standard polysomnographic criteria of REM sleep (Rechtschaffen and Kales 1968) are maintained. Lucid dreaming can be trained (LaBerge 1980), which makes this phenomenon a promising research topic despite its rarity in untrained subjects. Although REM sleep muscle atonia prevents overt motor behavior, lucid dreamers are able to communicate their state by predefined volitional eye movements clearly discernable in the electrooculogram (LaBerge et al. 1981). These volitional eye movements can serve as temporal markers, allowing fMRI analysis of lucidly experienced dream content. Dreamed clenching of the left and right hand could thereby be associated with neural activation of the contralateral motor cortices (Dresler et al. 2011).

Comparing lucid and non-lucid REM sleep, Edelman's distinction between primary and secondary or higher-order consciousness is of value, since the contrast between lucid and non-lucid dreaming mirrors the contrast between primary and higher-order consciousness (Dresler et al. 2009; Hobson 2009). In Edelman's (2003) words: "Higher-order consciousness allows its possessors to go beyond the limits of the remembered present of primary consciousness. An individual's past history, future plans, and consciousness of being conscious all become accessible." Lucid dreaming may be critical to fully understanding the neural correlates of higher-order consciousness, because in contrast to coma-wake and sleep-wake comparisons, there is no major shift in vigilance state as defined by formal criteria: lucid REM sleep still is REM sleep proper according to the classical Rechtschaffen and Kales (and new AASM) criteria. Lucid dreaming therefore provides the only phenomenon we know of that can contrast primary consciousness as experienced in dreams with full-blown higher-order consciousness (Dresler et al. 2009; Spoormaker et al. 2010a). Neural correlates of lucid dreaming therefore inform neurobiological approaches of consciousness research by revealing brain regions associated with higher-order consciousness. Quantitative EEG data show that lucid dreaming is related to an activation of the DLPFC (Voss et al. 2009), normally deactivated in REM sleep. fMRI data of lucid dreaming confirm these findings. In addition, the bilateral precuneus, parietal lobules, and occipito-temporal cortices activated strongly during lucid dreaming as compared to non-lucid REM sleep (Dresler et al. 2011).

A growing body of research has studied the brain in the resting state during wakefulness. One of the described resting state networks, the so-called default mode network, has often been named as a precursor to consciousness. However, as research has shown that this network continues to fluctuate in light sleep, and can even be detected in deep sleep (Sämann et al. 2011), coma (Boly et al. 2008), or in anesthesized monkeys (Vincent et al. 2007), its role has been redefined to subserve internal mentations associated e.g. with mental imagery, autobiographical memory, future envisioning and mind wandering (Buckner et al. 2008). This interpretation is strengthened by its pattern of anti-correlation with the so-called attention system, subserving external awareness (Fox and Raichle 2007). Interestingly, this anti-correlation disappears in deep sleep (Sämann et al. 2011). Recently another resting state network, the so-called frontoparietal control system, has been proposed to allow switching between the default mode network and its opposed attention system (Vincent et al. 2008). Due to its proposed role as a kind of meta-network, the fronto-parietal control system might be seen as a prime candidate for underlying processes of metacognition – and therefore the emergence of lucidity in dreams. fMRI data of lucid dreaming indeed show strong overlap with the fronto-parietal control system (Dresler et al. 2012).

More recent network approaches study the sleeping brain with graph-theoretical analyses of EEG-guided fMRI data of sleeping subjects (Spoormaker et al. 2010b; Larson-Prior et al. 2011). It was shown that the capacity of the brain to integrate information (Tononi 2004) is reduced in both light and deep sleep, however through different mechanisms. Large-scale functional brain networks in light sleep stage 1 were characterised by increased cortico-cortical connectivity but exclusion of a critical hub, the bilateral thalamus. In contrast, a breakdown of long-distance cortico-cortical connectivity occurred in deep sleep (Spoormaker et al. 2010b). An increased clustering of local nodes was also observed in deep sleep, illustrating network configuration (a change in 'small-worldness') throughout sleep. The thalamus has been identified as a core region of the neural network of (primary) consciousness in a number of studies applying anaesthesia (Alkire and Miller 2005), as it is known to function as a relay station in the neural cascade of (sensory) information processing. Of note, Wehrle et al. (2007) showed an increase in thalamo-cortical networking in phasic REM sleep, which may be interpreted as an indication of increased consciousness underlying the vivid and self-centered scenarios during dreaming. Indeed, prototypical dreaming has been proposed to be bound to phasic REM sleep, while tonic REM sleep rather seems to be associated with dreaming activity qualitatively similar to that of NREM sleep (Molinari and Foulkes 1969).

In conclusion, neuroimaging of sleep is a potent tool in dream research. It can provide us with remarkable insights into mechanisms underlying fluctuating self-awareness and consciousness during sleep, and can even allow us to differentiate specific 'dream' contents as shown in lucid dreamers. Today, the application of neuroimaging methods for sleep research is still hampered by the complex experimental procedure, which makes the collection of larger study samples a tedious process. As a few research groups have shown, these experimental drawbacks can be overcome, and the promise of combined EEG/fMRI is the combination of the high temporal resolution of EEG with the high spatial resolution of fMRI, such that investigations are not limited to cortical surface activity alone. Also due to the development of new analysis approaches, dream imaging is a quickly growing and highly promising field in sleep research that will offer new and deep insights into the dreaming brain.

References

Alkire, M. T., & Miller, J. (2005). General anesthesia and the neural correlates of consciousness. *Progress in Brain Research, 150*, 229–244.

Andrade, K. C., Spoormaker, V. I., Dresler, M., Wehrle, R., Holsboer, F., Sämann, P. G., & Czisch, M. (2011). Sleep spindles and hippocampal functional connectivity in human NREM sleep. *Journal of Neuroscience, 31*, 10331–10339.

Boly, M., Phillips, C., Tshibanda, L., Vanhaudenhuyse, A., Schabus, M., Dang-Vu, T. T., Moonen, G., Hustinx, R., Maquet, P., & Laureys, S. (2008). Intrinsic brain activity in altered states of consciousness: How conscious is the default mode of brain function? *Annals of the New York Academy of Sciences, 1129*, 119–129.

Born, A. P., Rostrup, E., Miranda, M. J., Larsson, H. B., & Lou, H. C. (2002). Visual cortex reactivity in sedated children examined with perfusion MRI (FAIR). *Magnetic Resonance Imaging, 20*, 199–205.

Braun, A. R., Balkin, T. J., Wesensten, N. J., Gwadry, F., Carson, R. E., Varga, M., Baldwin, P., Belenky, G., & Herscovitch, P. (1998). Dissociated pattern of activity in visual cortices and their projections during human rapid eye movement sleep. *Science, 279*, 91–95.

Buckner, R. L., Andrews-Hanna, J. R., & Schacter, D. L. (2008). The brain's default network. *Annals of the New York Academy of Sciences, 1124*, 1–38.

Caporro, M., Haneef, Z., Yeh, H. J., Lenartowicz, A., Buttinelli, C., Parvizi, J., & Stern, J. M. (2012). Functional MRI of sleep spindles and K-complexes. *Clinical Neurophysiology, 123*, 303–309.

Czisch, M., & Wehrle, R. (2010). Sleep. In C. Mulert & L. Lemieux (Eds.), *EEG–fMRI*. Berlin: Springer.

Czisch, M., Wetter, T. C., Kaufmann, C., Pollmächer, T., Holsboer, F., & Auer, D. P. (2002). Altered processing of acoustic stimuli during sleep: Reduced auditory activation and visual deactivation detected by a combined fMRI/EEG study. *NeuroImage, 16*, 251–258.

Czisch, M., Wehrle, R., Kaufmann, C., Wetter, T. C., Pollmächer, T., & Auer, D. P. (2004). Functional MRI during sleep: BOLD signal decreases and their electrophysiological correlates. *The European Journal of Neuroscience, 20*, 566–574.

Czisch, M., Wehrle, R., Stiegler, A., Peters, H., Andrade, K., Holsboer, F., & Sämann, P. G. (2009). Acoustic oddball during NREM sleep: A combined EEG/fMRI study. *PloS One, 4*, e6749.

Dang-Vu, T. T., Schabus, M., Desseilles, M., Albouy, G., Boly, M., Darsaud, A., Gais, S., Rauchs, G., Sterpenich, V., Vandewalle, G., Carrier, J., Moonen, G., Balteau, E., Degueldre, C., Luxen, A., Phillips, C., & Maquet, P. (2008). Spontaneous neural activity during human slow wave sleep. *Proceedings of the National Academy of Sciences, 105*, 15160–15165.

Desseilles, M., Dang-Vu, T. T., Sterpenich, V., & Schwartz, S. (2011). Cognitive and emotional processes during dreaming: A neuroimaging view. *Consciousness and Cognition, 20*(4), 998–1008.

Dresler, M., Wehrle, R., Spoormaker, V. I., Koch, S., Holsboer, F., Steiger, A., Obrig, H., Sämann, P. G., & Czisch, M. (2009). Neural correlates of consciousness – Insights from sleep imaging. *Neuroforum, 15*(S1), T24–3C.

Dresler, M., Koch, S. P., Wehrle, R., Spoormaker, V. I., Holsboer, F., Steiger, A., Sämann, P. G., Obrig, H., & Czisch, M. (2011). Dreamed movement elicits activation in the sensorimotor cortex. *Current Biology, 21*, 1–5.

Dresler, M., Wehrle, R., Spoormaker, V. I., Koch, S. P., Holsboer, F., Steiger, A., Obrig, H., Sämann, P. G., & Czisch, M. (2012). Neural correlates of dream lucidity obtained from contrasting lucid versus non-lucid REM sleep: A combined EEG/fMRI case study. *Sleep, 35*, 1017–1020.

Edelman, G. M. (2003). Naturalizing consciousness: A theoretical framework. *Proceedings of the National Academy of Sciences, 100*, 5520–5524.

Ermis, U., Krakow, K., & Voss, U. (2010). Arousal thresholds during human tonic and phasic REM sleep. *Journal of Sleep Research, 19*, 400–406.

Fox, M. D., & Raichle, M. E. (2007). Spontaneous fluctuations in brain activity observed with functional magnetic resonance imaging. *Nature Reviews Neuroscience, 8*, 700–711.

Hobson, J. A. (2009). The neurobiology of consciousness: Lucid dreaming wakes up. *International Journal of Dream Research, 2*, 41–44.

Hobson, J. A., & McCarley, R. W. (1977). The brain as a dream state generator: an activation-synthesis hypothesis of the dream process. *American Journal of Psychiatry, 134*, 1335–1348.

Hobson, J. A., & Pace-Schott, E. (2002). The cognitive neuroscience of sleep: Neuronal systems, consciousness and learning. *Nature Reviews Neuroscience, 3*, 679–693.

Hong, C. C., Harris, J. C., Pearlson, G. D., Kim, J., Calhoun, V. D., Fallon, J. H., Golay, X., Gillen, J. S., Simmonds, D. J., van Zijl, P. C. M., Zee, D. S., & Pekar, J. J. (2009). fMRI evidence for multisensory recruitment associated with rapid eye movements during sleep. *Human Brain Mapping, 30*, 1705–1722.

Jahnke, K., von Wegner, F., Morzelewski, A., Borisov, S., Maischein, M., Steinmetz, H., & Laufs, H. (2012). To wake or not to wake? The two-sided nature of the human K-complex. *NeuroImage, 59*, 1631–1638.

Kaufmann, C., Wehrle, R., Wetter, T. C., Holsboer, F., Auer, D. P., Pollmächer, T., & Czisch, M. (2006). Brain activation and hypothalamic functional connectivity during human NREM sleep: An EEG/fMRI study. *Brain, 129*, 655–667.

LaBerge, S. (1980). Lucid dreaming as a learnable skill: A case study. *Perceptual and Motor Skills, 51*, 1039–1042.

LaBerge, S., Nagel, L., Dement, W. C., & Zarcone, V. (1981). Lucid dreaming verified by volitional communication during REM sleep. *Perceptual and Motor Skills, 52*, 727–732.

Larson-Prior, L. J., Power, J. D., Vincent, J. L., Nolan, T. S., Coalson, R. S., Zempel, J., Snyder, A. Z., Schlaggar, B. L., Raichle, M. E., & Petersen, S. E. (2011). Modulation of the brain's functional network architecture in the transition from wake to sleep. *Progress in Brain Research, 193*, 277–294.

Maquet, P., Péters, J., Aerts, J., Delfiore, G., Degueldre, C., Luxen, A., & Franck, G. (1996). Functional neuroanatomy of human rapid-eye-movement sleep and dreaming. *Nature, 383*, 163–166.

Maquet, P., Degueldre, C., Delfiore, G., Aerts, J., Péters, J., Luxen, A., & Franck, G. (1997). Functional neuroanatomy of human slow wave sleep. *Journal of Neuroscience, 17*, 2807–2812.

Miyauchi, S., Misaki, M., Kan, S., Fukunaga, T., & Koike, T. (2009). Human brain activity time-locked to rapid eye movements during REM sleep. *Experimental Brain Research, 192*, 657–667.

Molinari, S., & Foulkes, D. (1969). Tonic and phasic events during sleep: Psychological correlates and implications. *Perceptual and Motor Skills, 29*, 343–368.

Nir, Y., & Tononi, G. (2009). Dreaming and the brain: From phenomenology to neurophysiology. *Trends in Cognitive Science, 14*, 88–100.

Rechtschaffen, A., & Kales, A. (1968). *A manual of standardized terminology, techniques and scoring system for sleep stages of human subjects*. Washington, DC: NIH.
Sämann, P. G., Wehrle, R., Hoehn, D., Spoormaker, V. I., Peters, H., Tully, C., Holsboer, F., & Czisch, M. (2011). Development of the brain's default mode network from wakefulness to slow wave sleep. *Cerebral Cortex, 21*(9), 2082–2093.
Schabus, M., Dang-Vu, T. T., Albouy, G., Balteau, E., Boly, M., Carrier, J., Darsaud, A., Degueldre, C., Desseilles, M., Gais, S., Phillips, C., Rauchs, G., Schnakers, C., Sterpenich, V., Vandewalle, G., Luxen, A., & Maquet, P. (2007). Hemodynamic cerebral correlates of sleep spindles during human non-rapid eye movement sleep. *Proceedings of the National Academy of Sciences, 104*, 13164–13169.
Schabus, M., Dang-Vu, T. T., Heib, D. P., Boly, M., Desseilles, M., Vandewalle, G., Schmidt, C., Albouy, G., Darsaud, A., Gais, S., Degueldre, C., Balteau, E., Phillips, C., Luxen, A., & Maquet, P. (2012). The fate of incoming stimuli during NREM sleep is determined by spindles and the phase of the slow oscillation. *Frontiers in Neurology, 3*, 40.
Spoormaker, V. I., Czisch, M., & Dresler, M. (2010a). Lucid and non-lucid dreaming: Thinking in networks. *International Journal of Dream Research, 3*, 49–51.
Spoormaker, V. I., Schröter, M. S., Gleiser, P. M., Andrade, K. C., Dresler, M., Wehrle, R., Sämann, P. G., & Czisch, M. (2010b). Development of a large-scale functional brain network during human non-rapid eye movement sleep. *Journal of Neuroscience, 30*, 11379–11387.
Tononi, G. (2004). An information integration theory of consciousness. *BMC Neuroscience, 5*, 42.
Vincent, J. L., Patel, G. H., Fox, M. D., Snyder, A. Z., Baker, J. T., Van Essen, D. C., Zempel, J. M., Snyder, L. H., Corbetta, M., & Raichle, M. E. (2007). Intrinsic functional architecture in the anaesthetized monkey brain. *Nature, 447*, 83–86.
Vincent, J. L., Kahn, I., Snyder, A. Z., Raichle, M. E., & Buckner, R. L. (2008). Evidence for a frontoparietal control system revealed by intrinsic functional connectivity. *Journal of Neurophysiology, 100*, 3328–3342.
Voss, U., Holzmann, R., Tuin, I., & Hobson, J. A. (2009). Lucid dreaming: A state of consciousness with features of both waking and non-lucid dreaming. *Sleep, 32*, 1191–1200.
Wehrle, R., Czisch, M., Kaufmann, C., Wetter, T. C., Holsboer, F., Auer, D. P., & Pollmächer, T. (2005). Rapid eye movement related brain activation in human REM sleep using fMRI. *Neuroreport, 16*, 853–857.
Wehrle, R., Kaufmann, C., Wetter, T. C., Holsboer, F., Auer, D., Pollmächer, T., & Czisch, M. (2007). Functional microstates within human REM sleep. *European Journal of Neuroscience, 25*, 863–871.

Chapter 14
Do You Think That Scientific Psychology Has a Place for the Study of Dreaming? In Other Words, Do You Accept Introspection as Scientifically Useful?

Michael W. Eysenck

Allan Hobson's lectures on Dream Consciousness represent an impressive contribution to our understanding of dreaming and consciousness, being both intellectually coherent and comprehensive. I agree with his central assumption that there are strong correlations between the brain and the mind. It is surely correct that identifying the precise pattern of brain activity during dreaming will provide important insights into the nature and function of consciousness.

I also agree with the notion that there is an important place for the study of dreaming within scientific psychology, and that introspection is scientifically useful. The optimal approach is to use converging operations to understand dreaming and consciousness. Self-report data, polygraphic data, and neuroimaging data possess different strengths and limitations, and so we need to compare and contrast the findings from all three approaches (Vogel 2000).

My third point of agreement is with the assumption that there are crucial similarities between dreaming and waking consciousness. This assumption is preferable to the Freudian notion that dream reports reveal crucial information about an individual that is inaccessible when the individual is awake.

My fourth point of agreement is that a major difference is that dreaming is largely limited to primary consciousness. In contrast, both primary and secondary consciousness can occur in the waking state. However, the difference should not be exaggerated given that dreams can involve higher-order cognitive activities such as focused attention and self-reflection (Kahan and Laberge 1996).

What are the points of disagreement? One relates to Hobson's implicit assumption that most dream reports are reliable and valid. The introspective evidence provided by dream reports is potentially valuable, but there are several caveats. There is an inevitable delay between having a dream and reporting it, and the ability to recall dreams decreases dramatically over time. Memory is state-dependent, and

M.W. Eysenck (✉)
Department of Psychology, Royal Holloway College, University of London, London, UK

Department of Psychology, University of Roehampton, London, UK
e-mail: m.eysenck@rhul.ac.uk

N. Tranquillo (ed.), *Dream Consciousness*, Vienna Circle Institute Library 3, DOI 10.1007/978-3-319-07296-8_14, © Springer International Publishing Switzerland 2014

the waking state in which dreams are reported differs considerably from the sleeping state in which dreams occur. It would be useful to have more research in which attempts were made to manipulate dream content experimentally.

In many studies, dream reports have been rated by independent judges. There are potential limitations with this approach. Dreamers reported that 40 % of their dream reports exhibited choice, whereas the figure was 10 % for judges rating the same dream reports (Kahan and Laberge 1996). More research needs to compare first-person and third-person ratings of dream content.

Hobson argues that dreams are triggered by quasi-random brain activity, are typically bizarre, often involve hallucinations and delusions, and can resemble psychotic thinking. These assumptions are of relevance to dreams spontaneously recalled upon waking in the morning. However, such dreams account for under 5 % of our dreams and are unrepresentative (Kahan and Laberge 1996). When dream reports are obtained in the laboratory by waking sleepers up during REM sleep, the great majority are mundane and organized rather than bizarre and random.

There are issues concerning the assumption that dreaming depends heavily on REM sleep and the pontine brainstem and so involves primarily bottom-up processing. Lesions that eliminate REM sleep rarely cause cessation of dreaming (Solms 2000). Brain-damaged patients who rarely or never dream nearly all have an intact pontine brain stem, but rather have damage to the forebrain around BA40 (Solms 2000). Since this area is associated with processes involving mental imagery, it may be preferable to argue that top-down processes determine dream content more than assumed by Hobson. We need to assess the detailed timing of processes in the brain during dreaming to clarify such causal issues.

The hypothesis that dreaming serves as a building block for waking consciousness is plausible. The connections between dreaming and waking consciousness are strengthened if we assume that the content of dreams is similar to (but less complex than) that of waking consciousness (Kahan and Laberge 1996) and that top-down processes play an important role in dreaming.

References

Kahan, T., & LaBerge, S. (1996). Cognition and metacognition in dreaming and waking. *Dreaming, 6*(2), 235–249.

Solms, M. (2000). Dreaming and REM sleep are controlled by different brain mechanisms. *Behavioral and Brain Sciences, 23*(6), 843–850.

Vogel, G. (2000). Critique of current dream theories. *Behavioral and Brain Sciences, 23*(6), 1014.

Chapter 15
What Does Your Theory of Hallucinosis Make of Dream Experience? Do Both Reveal the Operation of Internal Image Generator Mechanisms in the Brain?

Dominic H. Ffytche

Are dreams hallucinations? Hobson argues they are and that, as a result, each of us might be considered to have symptoms of mental illness several times a night. In support of Hobson's view, there is tell-tale acknowledgment of a problem in the diagnostic manual of mental symptoms (Diagnostic and Statistical Manual of Mental Disorders (DSM-IV)) (APA 1994):

> . . .The term hallucination is not ordinarily applied to the false perceptions that occur during dreaming, while falling asleep (hypnagogic), or when awaking (hypnopompic). . ..

From a clinical perspective, dreams are thus not hallucinations. Yet one might ask whether this arbitrary distinction hides a fundamental correspondence between the two. Is having a hallucination the same as having a dream in terms of subjective experience and underlying brain activity? Furthermore, one might ask where dreams and hallucinations sit in relation to other types of experience such as 'normal' perception (for vision, the day-to-day experience of seeing objects, spatial relations etc. that are present in the world around us i.e. 'true' as opposed to 'false' perceptions) and imagery, particularly as an important contemporary theory holds that hallucinations are simply misinterpreted imagery. If dreams and hallucinations are one and the same, based on this theory one would conclude that dreams are themselves a form of misinterpreted imagery. Here I attempt to answer these questions by focussing on particular aspects of dreamt, hallucinated, imagined and normally perceived experience – their phenomenology, neural mechanisms and visual content.

How do visual dreams and visual hallucinations compare in terms of their subjective qualities? The subjective experience of visually hallucinating or dreaming is to see things in external space, just as occurs during normal perception. Hallucinations and dreams have the same vivid quality as normal perception and we are not able to evoke or change them by force of will (Lucid dreams may be an

D.H. Ffytche (✉)
Department of Old Age Psychiatry, Institute of Psychiatry, King's College London, PO70, London, UK
e-mail: dominic.ffytche@kcl.ac.uk

N. Tranquillo (ed.), *Dream Consciousness*, Vienna Circle Institute Library 3,
DOI 10.1007/978-3-319-07296-8_15,

exception to this in that the dreamer has the impression they are controlling the dream plot. However it is possible the sense of volition is itself dreamt). These qualities distinguish them from visual imagery. Imagery refers to a specific cognitive operation, the experience you have when imagining Hobson's dream of riding a bicycle on a busy London street. Imagery is seen in a poorly defined internal space we refer to as the mind's eye and, for most of us, the experience is vague and lacks the perceptual clarity of hallucinations and dreams. Furthermore, imagery is evoked by a force of will, unlike hallucinations and dreams where we are as spectators observing events unfold. From the subjective viewpoint, visual dreams, visual hallucinations and normal visual perception are closely related but distinct from visual imagery (see Ffytche 2005, 2007).

The brain activity underlying these experiences leads us to the same conclusion. Normal visual percepts are associated with activation of occipital and temporal lobe regions specialised for different aspects of vision. A general principle seems to be that conscious percepts of a given visual attribute relate to activity within the cortex specialised for that attribute. Thus, visual percepts of colour are associated with activity within the cortex specialised for colour and so forth. The same brain regions are activated during visual hallucinations (Ffytche et al. 1998). Thus the colour-specialised cortex is activated during hallucinations of colour and so forth. In contrast, activity in the lateral frontal and parahippocampal cortex predominates during visual imagery, with weaker activation of visually-specialised occipital and temporal regions. REM sleep dreams fall somewhere between the two. In common with hallucinations and normal perception REM dreams have prominent activation of the visually-specialised cortex. In common with imagery they have prominent activation of the parahippocampal cortex; however, dreams differ from imagery in that the lateral frontal cortex is not activated. Thus while visual dreams, visual hallucinations, and normal visual perception are not identical in terms of underlying brain activity, they share prominent activation of the visually-specialised cortex. Activity in these visually-specialised regions seems to impart a perceptual quality to the experience and I refer to it as the neural basis of the 'eye's mind' contrasting with primarily the lateral frontal and parahippocampal activity underlying the 'mind's eye.' (Ffytche 2013)

Visual dreams are thus more like visual hallucinations and normal visual percepts than they are like visual imagery in terms of subjective experience and underlying brain activity; yet, can we say visual dreams and visual hallucinations are one and the same? Put another way, are visual hallucinations visual dreams appearing in the wake state and visual dreams visual hallucinations occurring during sleep? The answer is complicated by the fact that we now recognise more than one class of brain-mind activity meeting the definition of a visual hallucination. One class of brain-mind activity is associated with eye disease and is referred to as the Charles Bonnet Syndrome (CBS), after the philosopher/scientist who first described the condition (Ffytche 2005, 2007). CBS hallucinations range from simple shapes, lines, and coloured blobs to complex formed patterns and objects. Patterns include brickwork, tapestries, and lattices. Objects include faces, typically grotesque with prominent eyes and distorted features, figures, often in period costume like the 18th century characters in Hobson's bicycle dream, and extended landscape scenes. CBS hallucinations typically last seconds to minutes and while patients may be fooled initially, they invariably learn to recognise that the experiences are not real. The cause of CBS hallucinations seems to be cortical hypersensitivity within specialised visual regions resulting from loss of visual inputs to the brain from the eye. In terms of Hobson's AIM model, the visual brain is offline, just as occurs through gating during REM sleep. However, the fabric of our typical REM dreamscapes is not of coloured blobs, brickwork patterns, distorted faces, and eighteenth century figures, as might be expected if brain-mind activity in REM dreams were the same as in CBS hallucinations. In fact, recent work in our laboratory suggests that CBS is more closely related to the neurophysiology of NREM than REM sleep (Ffytche 2008) and that CBS hallucinations are best conceived as snapshots of the neural mechanisms of visual processing occurring outside their normal context (Ffytche 2010).

Another class of brain-mind activity associated with visual hallucinations is found in conditions such as Parkinson's disease, Alzheimer's disease, delirium, schizophrenia and brainstem lesions (peduncular hallucinations also referred to as L'Hermitte's syndrome (Ffytche 2007)). Unlike CBS, the visual pathways in these disorders are intact or minimally impaired and their associated visual hallucinations are very different to those described above. Peduncular hallucinations have a dream–like quality with an evolving narrative structure and multiple sensory modalities (e.g. figures that are seen, heard, and felt). Unlike CBS, penduncular hallucinations are believed real, both while they are occurring and after they have resolved. Visual hallucinations in Parkinson's disease, Alzheimer's disease, schizophrenia and delirium share many of these features and I refer to this class of brain-mind activity as the cholinergic syndrome to differentiate it from CBS. As in REM sleep, both the specialised visual cortex and parahippocampal cortex are activated during cholinergic syndrome hallucinations.

The list of distinct brain-mind activities linked to visual hallucinations goes on. A third syndrome is related to the serotonergic system and there are likely to be more. The deconstruction of visual hallucinations from a single class of brain-mind activity to a range of different brain-mind activities has implications for the relationship of hallucinations to dreams and imagery. Visual dreams are visual hallucinations but the converse is not also true: not all visual hallucinations are dreams. Visual hallucinations are best conceived as a disparate group of brain-mind activities arising in a variety of pathological and non-pathological contexts sharing the endpoint of conscious visual experience. It seems plausible that cholinergic syndrome hallucinations are REM fragments intruding into the wake state, but the same is not true of CBS and other types of hallucination. Dreams are thus one of several types of brain-mind activity subsumed under the term hallucination. Some types of hallucination may be related to misinterpreted imagery while others not and the place of dreams in this spectrum is unclear. Yet, from Hobson's perspective, the conceptual evolution of hallucinations from one to many classes of brain-mind activity merely expands his key point. By studying the neural mechanisms of dreams – or other types of hallucination – we gain insights into the neural mechanisms of consciousness.

References

American Psychiatric Association. (1994). *Diagnostic and statistical manual of mental disorders* (4th ed.). Washington, DC: Author.

Ffytche, D. H. (2005). Visual hallucinations and the Charles Bonnet Syndrome. *Current Psychiatry Reports, 7*(3), 168–179.

Ffytche, D. H. (2007). Visual hallucinatory syndromes: Past, present, and future. *Dialogues in Clinical Neuroscience, 9*(2), 173–189.

Ffytche, D. H. (2008). The hodology of hallucinations. *Cortex, 44*(8), 1067–1083.

Ffytche, D. H. (2010). The visual unconscious: Perspectives from the Charles Bonnet Syndrome. In E. Perry, D. Collerton, F. E. N. LeBeau, & H. Ashton (Eds.), *New horizons in the neuroscience of consciousness*. Amsterdam: John Benjamins.

Ffytche, D. H. (2013). The hallucinating brain: Neurobiological insights into the nature of hallucinations. In F. Macpherson & D. Platchias (Eds.), *Hallucination*. Cambridge, MA: MIT Press.

Ffytche, D. H., et al. (1998). The anatomy of conscious vision: An fMRI study of visual hallucinations. *Nature Neuroscience, 1*(8), 738–742.

Chapter 16
Please Comment on the Predictive Tenet of the Protoconsciousness Hypothesis. Is This Idea Consistent with the Helmholtzian Model of Free Energy That You Are Developing?

Karl Friston

16.1 Free-Energy and Sleep

16.1.1 Introduction

It is a great pleasure to comment upon Allan Hobson's theory of dreaming and protoconsciousness. When I first encountered these ideas (Hobson 2009), I was struck by the remarkable convergence between his analysis and the conclusions of a more technical treatment furnished by the free-energy principle (Friston 2010). In what follows, I try to emphasize the deep connections between these theories by considering some of Hobson's compelling observations, using the free-energy formulation.

16.1.2 Free-Energy and Generative Models

> REM sleep may constitute a protoconscious state, providing a virtual reality model of the world that is of functional use to the development and maintenance of waking consciousness. (Hobson 2009)

The two key observations here are that the brain possesses a model of the world and that REM sleep optimizes that model for waking exchanges with the environment. The first half of this commentary focuses on the necessary role of such models, while the second considers their optimization during sleep. To understand the relationship between Hobson's formulation and free-energy, we will rehearse the free-energy formulation briefly. It starts with a very simple premise and leads,

K. Friston (✉)
The Wellcome Trust Centre for Neuroimaging, Institute of Neurology, London, UK
e-mail: k.friston@fil.ion.ucl.ac.uk

N. Tranquillo (ed.), *Dream Consciousness*, Vienna Circle Institute Library 3,
DOI 10.1007/978-3-319-07296-8_16, © Springer International Publishing Switzerland 2014

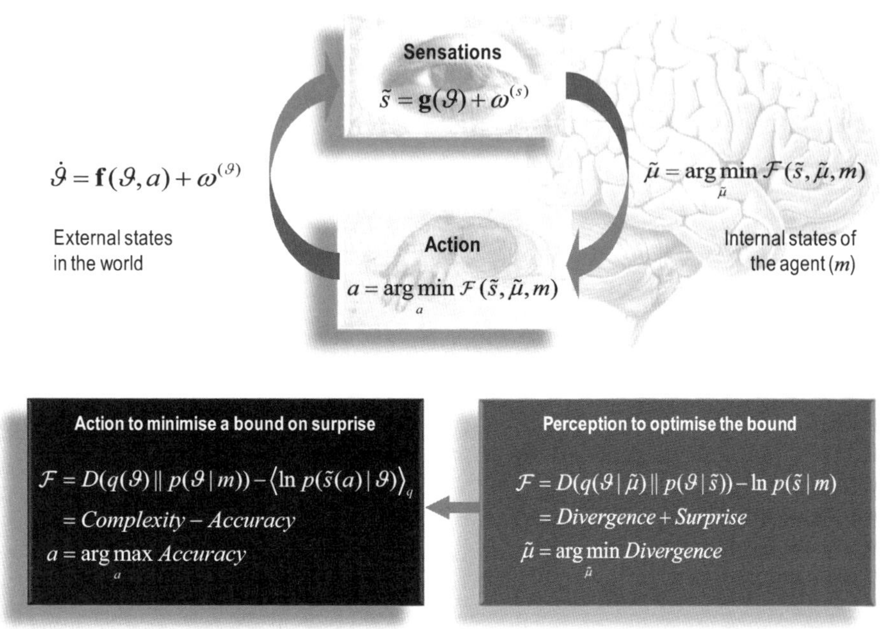

Fig. 16.1 The free-energy principle: This schematic shows the dependencies among quantities that define the free-energy of an agent or brain, denoted by m. These include its generalised internal states $\tilde{\mu}(t)$ and sensory signals $\tilde{s}(t)$. Generalised states include their generalised motion; velocity, acceleration etc. The environment is described by equations that specify the motion of its states $\vartheta(t)$, which depend on action $a(t)$. Both internal states and action minimise free-energy $\mathcal{F}(,\tilde{s},\tilde{\mu},m)$, which is a function of sensory input and the internal states. Internal states encode a proposal density $q(\vartheta|\tilde{\mu})$ on the causes of sensory input. These comprise states of the world and the amplitude of any random fluctuations $\omega(t)$. The *lower panels* provide the key equations behind the free-energy formulation. The *right* equality shows that optimizing internal states makes the proposal density an approximate conditional density on the causes of sensory input. Furthermore, it shows that free-energy is an upper bound on surprise. This is because the first term of the equality is a divergence between the proposal density and the true conditional or posterior density. Because this divergence can never be less than zero, minimizing free-energy renders it a proxy for surprise. At the same time, the proposal density becomes the posterior density. The *left* equality shows that action during wakefulness can only reduce free-energy by selectively sampling sensory data that are predicted under the proposal density. It also shows that, during sleep, complexity can be reduced by optimizing the model and its implicit prior $p(\vartheta|m)$. This entails making the prior more like the proposal density, to reduce their divergence

inescapably, to the notion that the brain *is* (sic) a generative model, whose optimization explains:

> its ability to create a remarkably reliable simulacrum of the world. (Hobson 2009)

In a nutshell, adaptive (self-organizing) agents resist a natural tendency to disorder (Ashby 1947; Nicolis and Prigogine 1977; Haken 1983). Mathematically, this requires them to minimise the average surprise or entropy of their sensations. To do this, they have to quantify the change in surprise due to their behavior

(Friston et al. 2010). However, the evaluation of surprise *per se* is mathematically intractable. This problem can be finessed by minimizing a quantity called free-energy (Feynman 1972), which, by construction, is always bigger than surprise (see Fig. 16.1).

$$\text{Free-energy} = \text{surprise} + \text{divergence between the proposal and posterior}$$

This construction uses a nonnegative divergence between a probabilistic representation of the causes of sensory input (a proposal density) and the true posterior density over these causes. Crucially, free-energy can be quantified given a model of the world that generates predictions of sensory input. In other words, converting surprise into free-energy induces a generative model of the sensorium. Minimising free-energy turns the proposal density into the posterior density over the causes of sensory input. This is the Bayesian brain hypothesis (Kersten et al. 2004) and casts the brain as an inference machine (von Helmholtz 1866; Dayan et al. 1995), where perception can be regarded as hypothesis testing (Gregory 1980):

> According to this hypothesis, our brains are as much creative artists as they are copy editors. (Hobson 2009)

There are many biologically plausible schemes that minimize free-energy, such as predictive coding (Mumford 1992; Rao and Ballard 1998; Friston 2010). However, our focus here is on the role of dreaming. How can one optimise a model in the absence of sensory information?

16.1.3 Dreaming and Complexity

> This turning on its head of the habitually reductive approach to dream interpretation sets the stage for a view of dreaming as an auto creative process. (Hobson 2009)

So what is created in the absence of new information? The answer lies in the nature of a good model, which (by appeal to Occam's principle) must balance its complexity and accuracy. This ensures our model of the world is veridical, in the sense that it is parsimonious and does not over-fit sensory data. This is a crucial part of the free-energy formulation, where we can rewrite free-energy as:

$$\text{Free-energy} = \text{complexity} - \text{accuracy}$$

Complexity is the divergence between the proposal density and prior beliefs about the causes of sensory input. This means that free-energy can be reduced by changing the model, which specifies the prior density. This means (literally) that yesterday's posteriors become tomorrow's priors. Complexity reduction eliminates redundant model parameters, accrued during wakefulness, to provide a more robust and simpler explanation for subsequent sensory encounters.

In machine learning there are mathematical precedents for this sort of model optimization (e.g., the wake-sleep algorithm; Hinton et al. 1995 and *post hoc* model optimization Friston and Penny 2011). Central to these schemes is the use of two phases, in which training proceeds with and without external sensory input. From our point of view, training without input (during sleep) corresponds to changing connections in the brain to minimise free-energy through minimizing complexity. Generally, complexity is reduced by eliminating redundant parameters (i.e., synaptic connections) to provide a simpler and sparser representation of causal structure in data. Although easy to do mathematically, the brain has to evaluate the relative free-energy of different models (priors) of the same sensory data. If the brain does this during sleep, there are two fundamental implications. First, there needs to be a replay of sensory inputs that are consistent with our model of the world, acquired during wakefulness. This may explain why:

> external inputs are . . . simulated in dreaming. I propose that the developing REM-sleeping brain has built-in predictions of external space and time. These intrinsic predictions are then adjusted on the basis of experiences of the outside world. (Hobson 2009)

Happily, fictive inputs are readily available because the brain can generate them using its generative model. Hobson considers some electrophysiological phenomena that might reflect this process:

> PGO waves may therefore constitute endogenously generated signals that . . . constitute informational building blocks for perception and motor control. During dreaming our sensation and movement are entirely fictive, but the simulation of real, wake-state sensation and movement is impressive. The brain, isolated from the outside world, treats this endogenous stimulation as if it were exogenous. (Hobson 2009)

These autopoietic sensations can be simulated easily in biologically plausible formulations of free-energy minimization. This involves suppressing the precision of sensory inputs (to emphasize top-down predictions), using exactly the same neuromodulatory mechanisms implicated in sleep maintenance. More generally, the implicit optimization of synaptic gain in the brain's hierarchical cortical models (Yu and Dayan 2005; Friston 2010) may speak to why it:

> possesses the means of regulating its own activation. (Hobson 2009)

In short, the brain may nuance and finesse its model of the world during sleep by finding simpler explanations for an internally generated sensorium. The second implication is that changing prior beliefs, to make them more like posterior beliefs, should not induce (fictive) experience-dependent learning or new memories.

16.1.4 Model Optimization Without Memories?

> Could it be that we must relearn all that we already know, as well as integrate new experiences into that vast storehouse of data? Or do we simply need to rerun our built-in virtual reality simulation circuits to prevent disuse-induced memory loss? (Hobson 2009)

The analysis above suggests both; in the sense that integrating new experiences into a model requires a parsimonious adjustment of prior beliefs entailed by that model. It suggests that the synaptic reorganization during sleep is concerned with eliminating redundancy (i.e., consolidation); not with establishing new memories. The implicit synaptic regression fits comfortably with the synaptic homoeostasis hypothesis (Gilestro et al. 2009), which suggests that synapses are lost during sleep. It also predicts that we will form no new fictive memories during sleep:

> If sleep is essential to memory, we must wonder why semantic memory does not seem to be strongly enhanced by sleep, why the enhancement of procedural learning by sleep, although statistically significant, is so weak. (Hobson 2009)

From the free-energy perspective, optimizing generative models in the brain through reducing their complexity is not about making them more accurate predictors but to consolidate and finesse predictions so that they generalize to contexts yet to be encountered. As noted by Hobson:

> All of these theories support the idea that sleep is as much a preparatory as it is a recovery process. (Hobson 2009)

16.1.5 Protoconsciousness and the Emergence (Modeling) of Self

> Originally dreamless, protoconscious REM sleep could therefore provide a virtual world model, complete with an emergent imaginary agent (the protoself) that moves (via fixed action patterns) through a fictive space (the internally engendered environment) and experiences strong emotion as it does so.

I was taken by this idea, particularly because it speaks to a great challenge for the free-energy formulation. Namely, how do models of self and others arise in (protoconscious) generative models of the world? This is clearly a key question that underlies theory of mind and awareness of self (Gallese and Goldman 1998; Frith and Frith 1999). It may be the case that the simplest explanation for our unsurprising and reproducible sensory interactions with the world is that they are caused by ourselves (and others); and that this hypothesis (protoself) emerges as a key component of generative models that provide simple (low free-energy) predictions of waking sensations.

References

Ashby, W. R. (1947). Principles of the self-organizing dynamic system. *Journal of General Psychology, 37*, 125–128.

Dayan, P., Hinton, G. E., & Neal, R. M. (1995). The Helmholtz machine. *Neural Computation, 7*, 889–904.

Feynman, R. P. (1972). *Statistical mechanics*. Reading: Benjamin.

Friston, K. (2010). The free-energy principle: A unified brain theory? *Nature Reviews Neuroscience, 11*(2), 127–138.
Friston, K., & Penny, W. (2011). Post hoc Bayesian model selection. *NeuroImage, 56*(4), 2089–2099.
Friston, K. J., Daunizeau, J., Kilner, J., & Kiebel, S. J. (2010). Action and behavior: A free-energy formulation. *Biological Cybernetics, 102*(3), 227–260.
Frith, C. D., & Frith, U. (1999). Interacting minds-a biological basis. *Science, 286*, 1692–1695.
Gallese, V., & Goldman, A. (1998). Mirror-neurons and the simulation theory of mind reading. *Trends in Cognitive Science, 2*, 493–501.
Gilestro, G. F., Tononi, G., & Cirelli, C. (2009). Widespread changes in synaptic markers as a function of sleep and wakefulness in Drosophila. *Science, 324*(5923), 109–112.
Gregory, R. L. (1980). Perceptions as hypotheses. *Philosophical Transactions Royal Society of London, B290*, 181–197.
Haken, H. (1983). *Synergetics: An introduction. Non-equilibrium phase transition and self-organization in physics, chemistry and biology* (3rd ed.). Berlin: Springer.
Hinton, G. E., Dayan, P., Frey, B. J., & Neal, R. M. (1995). The "wake-sleep" algorithm for unsupervised neural networks. *Science, 268*(5214), 1158–1161.
Hobson, J. A. (2009). REM sleep and dreaming: Towards a theory of protoconsciousness. *Nature Reviews Neuroscience, 10*(11), 803–813.
Kersten, D., Mamassian, P., & Yuille, A. (2004). Object perception as Bayesian inference. *Annual Review of Psychology, 55*, 271–304.
Mumford, D. (1992). On the computational architecture of the neocortex. II. The role of cortico-cortical loops. *Biological Cybernetics, 66*, 241–251.
Nicolis, G., & Prigogine, I. (1977). *Self-organization in non-equilibrium systems* (p. 24). New York: Wiley.
Rao, R. P., & Ballard, D. H. (1998). Predictive coding in the visual cortex: A functional interpretation of some extra-classical receptive field effects. *Nature Neuroscience, 2*, 79–87.
von Helmholtz, H. (1866/1962). *Concerning the perceptions in general, in treatise on physiological optics* (3rd ed.). New York: Dover.
Yu, A. J., & Dayan, P. (2005). Uncertainty, neuromodulation, and attention. *Neuron, 46*(4), 681–692.

Chapter 17
How Does the Protoconsciousness Concept of Dreaming Fit with Your Model of the Animal Mind? Do Dogs, Parrots, and Monkeys "Think" Without Words?

Ludwig Huber

17.1 Burdening with Consciousness

Animal cognition researchers are (mainly) concerned with behavior, which they use to infer the underlying mechanisms of physiological control. Of course, an animal's internal representations of the external world must have their physical basis in the brain, therefore understanding the neural basis of cognition must be an essential component of fully understanding cognition in animals (and humans). However, presently there is a basic problem in integrating neurobiological techniques and knowledge with behavioral cognitive studies. The problem is that our behavioral knowledge of cognitive capabilities, rudimentary though it may be in many respects, is still more advanced than our neurobiological knowledge. Therefore, theories of cognition in non-human animals are mainly *functional*, lacking any deeper knowledge of the exact *content* of the stored representations. This is true for both consciousness and proto-consciousness. Dreaming is no exception. From the behavior of a sleeping dog, we may infer it is dreaming. For example, I once read the following web entry (http://petshub.com/dog/dog-dreaming.php):

> You can tell when your dog is dreaming by watching him. When he first falls asleep, his breathing will become more regular as the sleep becomes deeper. When the dream starts, the dog's breathing becomes shallow and irregular. There may be occasional muscle twitches, and you can actually see his eyes moving behind his closed lids if you look closely enough. The eyes are moving because the dog is actually looking at the dream images in the same way that he would look at real objects in the outside world.

But such reports cannot fool us that we are (currently?) unable to assess the content of the dream. Non-verbal animals cannot report about their dreams (maybe to conspecifics, but not to humans). And there is a further fundamental problem that confronts the study of cognitive processes in a comparative framework, called the

L. Huber (✉)
Messerli Research Institute, University of Veterinary Medicine, Vienna, Austria
e-mail: ludwig.huber@univie.ac.at

N. Tranquillo (ed.), *Dream Consciousness*, Vienna Circle Institute Library 3,
DOI 10.1007/978-3-319-07296-8_17,

learning-performance problem (Kamil 1998). That is, although we are interested in measuring learning abilities or cognitive abilities in our experiments, all we are ever able to measure directly is behavior, and the behavior we observe is a function of many factors besides cognitive ability. Therefore, whenever we obtain a difference between species in their performance on a cognitive task, how can we be sure that the difference truly represents a difference in cognitive ability?

Hobson likes the distinction between primary and secondary consciousness, with the latter depending on the evolution of language. While we humans have language and can give verbal reports of our states, "subhuman" animals are stuck with purely perceptional and emotional (proto-conscious) states. First of all, I would recommend to use the word "non-human" instead of "sub-human" to classify animals properly, avoiding a scholastic, anagenetic or teleological conception of evolution ("from amoeba to man"). It is an old philosophical question if thinking without language qualifies as thinking at all. The *anthropocentric* view considers the way we think as the prototypical (or essential or only) way of thinking. Our (propositional) thought is obviously related to language, which tempted many philosophers to believe that language and its resulting abilities is all that matters (Davidson 1984). Must therefore any other species also have language in order to think? What are distinctive capacities of thinking (or reasoning or rational) species? Two primary candidates in the contemporary philosophy of mind are the ability to generalize in an abstract way and to make mistakes (Hurley and Nudds 2006a, b). So we may ask if an animal must have concepts and conceptual abilities or language in order to generalize and to make (and recognize) mistakes.

As Herrnstein (Herrnstein and Loveland 1964) and many followers (Huber and Aust 2006) since the seminal work with pigeons have shown, categorization and concept discrimination do not depend on linguistic abilities. Nor do abstract and relational categorization depend on linguistic abilities (Huber 2010). Moreover, language is not an all-or-nothing ability, but has evolved gradually. There is sufficient evidence to suggest a distinction between the faculty of language in the broad and the narrow sense (Hauser et al. 2002). The first includes a sensory-motor system, a conceptual-intentional system, and the computational mechanisms for recursion, providing the capacity to generate an infinite range of expressions from a finite set of elements. According to this distinction, language in the narrow sense only includes recursion and is the only uniquely human component of the faculty of language. This faculty may have evolved for reasons other than language. Comparative studies are currently looking for evidence of such computations outside of the domain of communication (for example: number, navigation, and social relations) (Fitch et al. 2010).

What's about *displacement*? Can a non-verbal animal de-center from itself and take account of the perspective of another animal by simulation, by using its own practical abilities off-line? Or does such de-centering require theorizing about the other's mind and so require having both concepts of, and a theory about, mental states? If so, what mental states? And what qualifies as a mental state? The perception and knowledge of others? The goals and intentions of others? There is solid evidence from several different experimental paradigms that chimpanzees understand all of them (Tomasello et al. 2003). The same is true for some corvids, like ravens (Bugnyar and Heinrich 2005) and scrub jays (Clayton et al. 2007), when solving "knower-guesser" tasks: they can take others' perception into account and draw inferences about the probability of winning food from, or losing it to those others. But there is currently no evidence that chimpanzees or any other species understand *false beliefs*. Call and Tomasello, therefore, conclude that chimpanzees understand others in terms of a perception – goal psychology, as opposed to a full-fledged, human-like belief–desire psychology (Call and Tomasello 2008).

What about making mistakes? Must an animal have the concept of a mistake or of a reason in order to recognize that it has made or is likely to make a mistake – or could this again be registered and manifested in practical terms? Do animals know that they could be wrong? (Call 2010) A challenge for future animal research in the absence of the fine-grained distinctions that verbal reports make available is to design proper behavioral experiments to assess the capacities for de-centering or recognizing mistakes. What kinds of non-verbal behavior, and under what conditions, might provide evidence for these capacities?

A prominent feature of human awareness is that we know or can estimate the state of our memories, that is, assess our own memory strength. We are consciously aware of some memories and can make verbal reports about them. This "knowing what we know" ability has been called metamemory or metacognition (Flavell and Wellman 1977; Flavell 1979, 1992). It can function to guide efficient behaviour because it enables us to cope with uncertainty by escaping a task or seeking further information. Thus, metamemory as a kind of self-knowledge is adaptive and might be expected to be phylogenetically widespread. Respective comparative research may be useful in revealing similarities between animal and human cognition, in suggesting the precursors of human cognitive sophistication, and in pinpointing the special allowances of language for cognition. However, many human uncertainty paradigms do not support this comparison, for they rely on introspection and self-reports (feelings of knowing, tip-of-the-tongue experiences, etc.) and do not suit animal observers.

Recent years have indeed seen a considerable increase in interest in metacognitive abilities in nonhuman species, using a perceptual and behavioral uncertainty paradigm that could be used comparatively (Carruthers 2009). With animals, the best researchers can do is to look for nonverbal behavior that is *functionally* similar, i.e., the animals are asked to use a *behavioral proxy* for human uncertainty. Animals able to discern the presence and absence of memory should improve accuracy if allowed to decline memory tests when they have forgotten, and should decline tests most frequently when memory is attenuated experimentally. This can be done by providing the subjects with an *escape* response option. Choice of this option leads to a mediocre reward, less than can be obtained by completing the test correctly, but more than is available if the test is taken and failed. Thus, there is an incentive for the animal to use metamemory in order to maximize the reward received. Subjects that know how well they will perform on the primary task in a particular trial should selectively escape from difficult trials when allowed to, and when they choose to complete the primary task instead of opting out, they should perform better than when forced to complete it.

Metamemory has been established in dolphins (Smith et al. 1995), rhesus macaques (Hampton 2001; Hampton et al. 2004; Washburn et al. 2006; Son and Kornell 2005), capuchin monkeys (Basile et al. 2009; Beran et al. 2009; Fujita 2009), orangutans (Suda-King 2008) and even rats (Foote and Crystal 2007) by converging evidence from several paradigms. In contrast, pigeons have, so far, performed differently from mammals in tests on metamemory (Inman and Shettleworth 1999; Sole et al. 2003; Sutton and Shettleworth 2008; Roberts et al. 2009). Although they behaved adaptively in that they maximized perceived reward, they apparently did so without relying on a process having the functional properties of conscious uncertainty. However, metamemory in pigeons has only to be tested in only one (difficult) task (delayed matching to sample) and has never been explored in the context of long-term retention.

Humans comply with uncertainty by often reporting that choices are based on a strategic cognitive process aided by conscious self-regulation. This fits into the early psychologists' suggestion that cognitive conflict, difficulty, and confusion elicit higher modes of cognition and that "*consciousness provides extraneous help to cognition when nerve processes are hesitant*" (James 1890/1952). However, a more parsimonious interpretation of such metacognitive responses, in both humans and animals, is that they reflect controlled decisional and criterion-setting processes that are necessary in the perceptual ambiguity of threshold. "*This grants escape responses a cognitive sophistication that fits the data, without burdening them with consciousness or with equally heavy ad hoc behaviorist assumptions*" (Smith et al. 1997).

There is a risk in explaining animals' performance in human-designed tasks in the same way we would explain the behavior of human participants – say, in terms of processes of inference and reasoning, or metacognition, or mind reading. For instance, animal subjects in metacognitive experiments involving an escape response may react to uncertainty without being aware of their knowledge states or their level of uncertainty (Staddon et al. 2007; Carruthers 2009). Or they may simply react to the anxiety that they may feel in uncertain situations without representing their state of anxiety. No doubt, *different* processes can lead to *similar* behavior (even among human beings, for example, in recovery of function after brain damage). "*Without language, how can we determine whether to attribute a rational process to explain an animal's behavior? More fundamentally, what kinds of processes are the 'right kind' to count as rational?*" (Hurley and Nudds 2006a, b)

References

Basile, B. M., Hampton, R. R., Suomi, S. J., & Murray, E. A. (2009). An assessment of memory awareness in tufted capuchin monkeys (Cebus apella). *Animal Cognition, 12*, 169–180.

Beran, M. J., Smith, J. D., Coutinho, M. V., & Boomer, J. (2009). The psychological organization of "uncertainty" responses and "middle" responses: A dissociation in capuchin monkeys (Cebus apella). *Journal of Experimental Psychology: Animal Behavior Processes, 35*(3), 371–381.

Bugnyar, T., & Heinrich, B. (2005). Ravens, Corvus corax, differentiate between knowledgeable and ignorant competitors. *Proceedings of the Royal Society B: Biological Sciences, 272*(1573), 1641–1646.
Call, J. (2010). Do apes know that they could be wrong? *Animal Cognition, 13*, 689–700.
Call, J., & Tomasello, M. (2008). Does the chimpanzee have a theory of mind? 30 years later. *Trends in Cognitive Sciences, 12*, 187–192.
Carruthers, P. (2009). Meta-cognition in animals: A sceptical look. *Mind and Language, 23*, 58–89.
Clayton, N. S., Dally, J. M., & Emery, N. J. (2007). Social cognition by food-caching corvids. The western scrub-jay as a natural psychologist. *Philosophical Transactions of the Royal Society B: Biological Sciences, 362*(1480), 507–522.
Davidson, D. (1984). Thought and talk. In: *Essays on truth and interpretation* (pp 155–170). Oxford: Clarendon Press.
Fitch, W. T., Huber, L., & Bugnyar, T. (2010). Social cognition and the evolution of language: Constructing cognitive phylogenies. *Neuron, 65*(6), 795–814.
Flavell, J. H. (1992). Perspectives on perspective taking. In H. Beilin & P. Pufall (Eds.), *Piaget's theory: Prospects and possibilities* (pp. 107–139). Hillsdale: Erlbaum.
Flavell, J. H. (1979). Metacognition and cognitive monitoring: A new area of cognitive-developmental inquiry. *American Psychologist, 34*, 906–911.
Flavell, J. H., & Wellman, H. M. (1977). Metamemory. In R. V. Kail Jr. & J. W. Hagen (Eds.), *Perspectives on the development of memory and cognition* (pp. 3–33). Hillsdale: Erlbaum.
Foote, A. L., & Crystal, J. D. (2007). Metacognition in the rat. *Current Biology, 17*, 551–555.
Fujita, K. (2009). Metamemory in tufted capuchin monkeys (Cebus apella). *Animal Cognition, 12*, 575–585.
Hampton, R. R. (2001). Rhesus monkeys know when they remember. *Proceedings of the National Academy of Sciences, 98*, 5359–5362.
Hampton, R. R., Zivin, A., & Murray, E. A. (2004). Rhesus monkeys (Macaca mulatta) discriminate between knowing and not knowing and collect information as needed before acting. *Animal Cognition, 7*, 239–246.
Hauser, M. D., Chomsky, N., & Fitch, W. T. (2002). The faculty of language: what is it, who has it, and how did it evolve? *Science, 298*, 1569–1579.
Herrnstein, R. J., & Loveland, D. H. (1964). Complex visual concept in the pigeon. *Science, 146*, 549–551.
Huber, L. (2010). Categories and concepts: Language-related competences in non-linguistic species. In M. D. Breed & J. Moore (Eds.), *Encyclopedia of animal behavior* (pp. 261–266). Oxford: Academic.
Huber, L., & Aust, U. (2006). A modified feature theory as an account of pigeon visual categorization. In E. A. Wasserman & T. R. Zentall (Eds.), *Comparative cognition: Experimental explorations of animal intelligence* (pp. 325–342). New York: Oxford University Press.
Hurley, S., & Nudds, M. (Eds.). (2006a). *Rational animals?* Oxford: Oxford University Press.
Hurley, S., & Nudds, M. (2006b). The questions of animal rationality: Theory and evidence. In: *Rational animals?* (pp 1–83). Oxford: Oxford University Press.
Inman, A., & Shettleworth, S. J. (1999). Detecting metamemory in nonverbal subjects: A test with pigeons. *Journal of Experimental Psychology: Animal Behavior Processes, 25*, 389–395.
James, W. (1890/1952). *The principles of psychology: Vol. 53. Great books of the Western world* (p. 93). Chicago: University of Chicago Press.
Kamil, A. C. (1998). On the proper definition of cognitive ethology. In R. P. Balda, I. M. Pepperberg, & A. C. Kamil (Eds.), *Animal cognition in nature: the convergence of psychology and biology in laboratory and field* (pp. 1–28). San Diego: Academic.
Roberts, W. A., Feeney, M. C., McMillan, N., MacPherson, K., Musolino, E., & Petter, M. (2009). Do pigeons (Columba livia) study for a test? *Journal of Experimental Psychology: Animal Behavior Processes, 35*(2), 129–142.

Smith, J. D., Schull, J., Strote, J., McGee, K., Egnor, R., & Erb, L. (1995). The uncertain response in the bottlenosed dolphin (Tursiops truncatus). *Journal of Experimental Psychology: General, 124*, 391–408.

Smith, J. D., Shields, W. E., Schull, J., & Washburn, D. A. (1997). The uncertain response in humans and animals. *Cognition, 62*, 75–97.

Sole, L. M., Shettleworth, S. J., & Bennett, P. J. (2003). Uncertainty in pigeons. *Psychonomic Bulletin & Review, 10*(3), 738–745.

Son, L., & Kornell, N. (2005). Meta-confidence judgments in rhesus macaques: Explicit versus implicit mechanisms. In H. Terrace & J. Metcalfe (Eds.), *The missing link in cognition: Origins of self-reflective consciousness* (pp. 296–320). Oxford: Oxford University Press.

Staddon, J. E. R., Jozefowiez, J., & Cerutti, D. (2007). Metacognition: A problem not a process. *PsyCrit,* 1–5.

Suda-King, C. (2008). Do orangutans (Pongo pygmaeus) know when they do not remember? *Animal Cognition, 11*, 21–42.

Sutton, J., & Shettleworth, S. J. (2008). Memory without awareness: Pigeons do not show metamemory in delayed matching to sample. *Journal of Experimental Psychology: Animal Behavior Processes, 34*(2), 256–282.

Tomasello, M., Call, J., & Hare, B. (2003). Chimpanzees understand psychological states – The question is which ones and to what extent. *Trends in Cognitive Sciences, 7*, 153–156.

Washburn, D. A., Smith, J. D., & Shields, W. E. (2006). Rhesus monkeys (Macaca mulatta) immediately generalize the uncertain response. *Journal of Experimental Psychology: Animal Behavior Processes, 32*, 185–189.

Chapter 18
How Does the New Protoconsciousness Hypothesis Fit with Your Own Concept of the Cognitive Unconscious?

John F. Kihlstrom

Philosophers who try to define consciousness often contrast it to a state of dreamless sleep. So I think that Allan Hobson is right to think that understanding sleep, and dreams, may be a key to that Holy Grail of psychological research – understanding consciousness itself, including its biological substrates. Hobson argues that dreaming is not an unconscious process but rather an altered state of consciousness which is poorly remembered – if it is remembered at all – in normal waking life. I agree, but it is one thing for dreaming to result from an unconscious *process* and another thing for dreaming to be an unconscious *mental state* (Kihlstrom 2010).

You do not have to be Freud to believe that dreaming results from unconscious processes that are executed involuntarily and unavailable to introspective phenomenal awareness. Hobson's AIM model hypothesizes that dreaming is just the adventitious result of certain patterns of physiological activity that occur during Stage REM: high levels of cortical activation plus the gating of external inputs and motor outputs plus a modulatory imbalance favoring cholinergic over aminergic activity equals dreaming – and it all just happens, automatically and unconsciously, as a consequence of the physiology of the sleep cycle. A psychological consequence of these physiological changes, apparently, is the more-or-less random activation of neural networks that correspond to various objects of cognition – a bicycle, London, a tree that needs trimming, etc. These are then synthesized into a makeshift narrative – riding a bicycle through London, on my way to trimming a tree.

But when does this synthesis happen? Are we actually conscious of our dreams as they occur during Stage REM? Or are we conscious of the dream only when we awaken from a dream and retrieve it from working memory – or perhaps in the ascension from Stage REM, as we're recovering our wits, but before completely awakening? Consider the implications of Maury's "Dream of the Guillotine," as recounted by Freud in Chapter 1 of *The Interpretation of Dreams* (1900). It seems

J.F. Kihlstrom (✉)
Department of Psychology, University of California, Berkeley, Berkeley, CA, USA
e-mail: jfkihlstrom@berkeley.edu

N. Tranquillo (ed.), *Dream Consciousness*, Vienna Circle Institute Library 3, DOI 10.1007/978-3-319-07296-8_18,

unlikely that Maury experienced an extended narrative dream in which the fall of the blade occurred precisely at the moment that his headboard fell on his neck. Rather (and this thought is not original with me), it seems more likely that the fallen headboard awakened Maury, and that the dream itself was a retrospective reconstruction, piecing together whatever happened to be in working memory at the time he awakened – including the sensation of the falling headboard. Put another way, perhaps Maury had no conscious awareness of whatever was going on during the REM period itself. Perhaps conscious awareness began only when he woke up, and when he woke up he pulled together the bits and pieces left active in working memory to construct the dream. Perhaps this constructive activity is responsible for the peculiar formal qualities of dream-mentation noted by Hobson.

The hypothesis that people are not aware of dreams as they occur, but only retrospectively, upon awakening – and that what they are aware of is a retrospective reconstruction, not the dream itself – would seem to be contradicted by the phenomenon of lucid dreaming. Lucid dreamers do appear to be conscious during their dreams: they are conscious of their dreams, and that their dreams are dreams, and not actual experiences. On the other hand, lucid dreaming appears to be "rare and evanescent" (Lecture I), unless subjects are expressly trained to do it, and even extensive training does not always succeed. Moreover, it is not at all clear that lucid dreaming is representative of ordinary dreaming. It may constitute a separate state of consciousness altogether – produced, if Hobson is right, by a loosening of the gating mechanism, and the restoration of some balance between aminergic and cholinergic activity (in other words, the sleeper is waking up, but hasn't quite gotten there yet).

As Hobson defines it, "protoconsciousness" in dreams resembles the cognitive unconscious because whatever ideas and images are activated during REM (e.g., representations of bicycles, London, and trees to be trimmed) would be available to consciousness (if we were awake), even if they are not accessible to phenomenal awareness. Something then acts to bring these mental elements into awareness, where we can monitor and control them and put them together into some sort of image or narrative. But the cognitive unconscious is more than just a collection of latent mental contents, available but not accessible to conscious awareness, and waiting to be activated. What makes implicit memory, implicit perception, and the like interesting for psychological theory is that the percepts and memories in question influence ongoing experience, thought, and action, as in priming effects, in the absence of phenomenal awareness. They are dynamically active – even if this is not the kind of dynamic activity envisaged by Freudian psychoanalysis.

The late patient H.M. famously described his amnesia as like "awakening from a dream" (Milner et al. 1968). But H.M. was clearly conscious of events as he experienced them. They were subsequently lost to conscious recollection, or explicit memory, but their traces were nevertheless visible in priming effects and other manifestations of implicit, or unconscious memory. Which begs the following questions: How certain can we be that we are conscious of dreams as they're generated by the AIM process, every 90 min or so, four or five times a night (depending on how much sleep you get)? If we are conscious of these dreams while they occur, why are they forgotten so soon afterward? And regardless of whether we are conscious of the dreams as they occur, do the activated ideas and images influence our conscious experience, thought, and action the way implicit percepts and memories do? How would we know?

Dreaming may well be an altered state of consciousness. But if there is no consciousness during the dream itself, and the manifest contents of a dream do not dynamically influence experience, thought, and action outside of phenomenal awareness and voluntary control, then the idea of dreaming as a state of "protoconsciousness" may be something of a misnomer. This is because whatever ideas and images are activated during Stage REM may be entirely unconscious – and when we awaken, all we are aware of are whatever traces happen to remain in working memory, H.M.-like – soon to disappear entirely unless these fragments are reconstructed into a dream experience.

References

Freud, S. (1900). *The interpretation of dreams* (Translated from German and edited by J. Strachey). New York: Basic Books.

Kihlstrom, J. F. (2010). Unconscious processes. In D. Reisberg (Ed.), *Oxford handbook cognitive psychology*. Oxford: Oxford University Press.

Milner, B., Corkin, S., & Teuber, H. L. (1968). Further analysis of the Hippocampal Amnesic syndrome: 14-year follow-up study of H. M. *Neuropsychologia, 6*(3), 215–234.

Chapter 19
Does Morrisson's PGO Wave/Startle Hypothesis Help Us Explain Such Robust Dream Features as Surprise and Scene Shift?

Don Kuiken

It is often difficult to respond to Allan Hobson's discussions of dreaming; so many issues are raised in his presentations that it is difficult to know where to begin. A further complication is that he, as a neuroscientist, and I, as a psychologist, are informed by different traditions of research. Perhaps for that reason, in our attempts to address aspects of dream research of which we are both apprised, our readings often are at odds. Sometimes, I believe, he reads through a distorting lens. For example, he argues that the peak date of current event incorporation into dreams is 6 days after the event, but actually there are two roughly comparable peaks, the first occurring within 1 or 2 days of the event (the day residue) and another occurring about 6 days later (the so-called dream lag effect). The psychoanalytic discussion of day residue is challenged but not overturned by the results of these contemporary dream studies.

However, I want to focus on a more ambiguous case in which Hobson's reading of the research is accurate as far as it goes—but incomplete. I have in mind Morrison's studies of cats in which lesions of the pontine tegmentum in the vicinity of the locus coeruleus produced REM sleep without atonia. According to Hobson, the cats that were in REM with ponto-geniculo-occipital (PGO) waves "jumped to their feet and executed a complex motor sequence of attack and defense postures." He continues:

> The readout of these behaviors during REM sleep-without-atonia would seem to suggest that the brain is automatically programmed to emit important survival movements. These movements may normally be inhibited but can be experienced by us in dream consciousness when we flee imaginary attackers or turn and confront our imaginary aggressors...

Although the results reported are compatible with the preceding conclusion, Hobson's description of Morrison's studies elides other important observations. In addition to attack and defense postures, the lesioned cats' PGO waves were

D. Kuiken (✉)
Department of Psychology, University of Alberta, Edmonton, AB, Canada
e-mail: dkuiken@ualberta.ca

N. Tranquillo (ed.), *Dream Consciousness*, Vienna Circle Institute Library 3, DOI 10.1007/978-3-319-07296-8_19,

accompanied by a variety of overt orienting activities (e.g., exploring, staring, searching). Moreover, in related studies, Morrison and his colleagues found that PGO waves also accompanied overt orienting activities during waking. This correspondence motivated them to consider PGO waves during either REM or waking as an index of the orienting response, i.e., the reflexive interruption of ongoing behaviour (a slight pause or hesitation) and the initiation of investigative behaviour (attentional redirection) that characterize reactions to unexpected stimuli. Because PGO waves elicited by unexpected stimuli during REM sleep habituate more slowly than during slow wave sleep, Morrison concluded that REM sleep is characterized by persistent and heightened activation of the orienting response.

This hypothesis has important implications for understanding the supposed "bizarreness" of REM dreaming. The orienting response is an adjustment to a discrepancy between a presented stimulus and the kind of stimulus for which the individual is prepared. When such a discrepancy occurs during wakefulness, the orienting process provides adjustments in working memory that prepare the individual for recognition of the unexpected stimulus. During REM sleep, the orienting response is endogenously induced and no external stimuli actually are registered because of the sensory blockade associated with sleep. Nonetheless, the dreaming individual plausibly reacts *as though* stimulus change has occurred, and such adjustments in working memory may be experienced as discontinuities in dream imagery. It is not surprising then, that in human REM sleep, PGO-generated peripheral phasic events (e.g., periorbital integrated potentials (PIPs) and middle ear muscle activity (MEMAs)) predict the initiation of such mnemonic adjustments, resulting in dream reports that include "bizarre" discontinuities.

Hobson's account of such "bizarreness" during dreaming is that REM sleep temporarily introduces "chaotic" pontine-generated activation into the cerebral network and prompts transition to a completely unrelated output pattern. By contrast, conceiving PGO waves during REM sleep as pontine-generated orienting activity suggests that attentional adjustments during dreaming do not produce completely unrelated output patterns. Rather, they produce output patterns that are conceptual neighbours of the outputs that preceded PGO activation. Roughly, the orienting response initiates changes in working memory in response to the [implicit] question, "What is this?" and the [implicit] answer is, "Oh, not [the expected] X, but [the unexpected] X." For example, thoughts of "aphids on the leaves of my tomatoes" may be transformed into imagery of "spiders on the leaves of my tomatoes." Or, if the orienting response is intense, relatively remote conceptual neighbours may be activated. For example, thoughts of "aphids on the leaves of my tomatoes" may be "metaphorically" transformed into images of "spiders on my hands." What is new in these insect-appendage patterns is only partly new, a variation on an already available theme—much as might be expected if dreaming were a metaphoric, or at least quasi-metaphoric, mode of thought.

The process that follows from the orienting response hypothesis is in marked contrast to Hobson's associative "filling in" hypothesis. Rather than an attempt to complete an expectation, the orienting response hypothesis suggests that dream discontinuities are an attempt to adjust to a disconfirmed expectation. By implication, the conceptual neighbours that emerge during dreaming are anomalous (as metaphors also often are) but they are not deficient (any more than metaphors often are). In this regard, Hobson's argument that dreaming is accompanied by de-activation of the dorsolateral prefrontal cortex (DLPFC) may be pivotal. He construes such de-activation as a deficit: de-activation of the DLPFC is isomorphic with the absence of reflective awareness, planning, and volition. However, evidence increasingly suggests that the DLPFC supports a quite specific temporally extended and planful form of self regulation. In contrast, Braun and his colleagues have provided evidence that a quite different kind of self-regulation, the listening for "what comes" characteristic of musical improvisation, is supported during waking by the same pattern of activations and de-activations (including de-activation of the DLPFC) that characterize REM sleep (Limb and Braun 2008). Thus, it remains plausible that the *abandonment* of a monitored, deliberate, and planful form of self-regulation during REM dreaming enables the *adoption* of an attentive, spontaneous, and fluid form of self-regulation of the kind that occurs during musical improvisation. Dreaming, in this view, entails self-regulated but fluid openness to conceptual neighbours, rather than the absence of direct self-monitoring and response inhibition that enable temporally extended and planful decision making. Dreaming may, in other words, entail the kind of self-regulated openness to unbidden conceptual neighbours that motivated Bert States to call dreaming "involuntary poetry."

The orienting response model prompts reconsideration of a number of Hobson's other hypotheses about dreaming. Much more careful attention must be given to changes in the functions of working memory during REM sleep. Much more careful consideration must be given to the differentiation between responses to unexpected events, especially between the whole-body disorientation that characterizes startle and the conceptual reorientation that characterizes the orienting response. And, much more careful consideration must be given to the difference between a conception of dream "creativity" that depends on the generation of novelty and a conception that depends upon the transformation of categorical thinking. The outcome of such efforts almost certainly will be a conception of dreaming that strikes a more balanced pose somewhere between disciplined rationality and disoriented madness.

Reference

Limb, C. J., & Braun, A. R. (2008). Neural substrates of spontaneous musical performance: An fMRI study of jazz improvisation. *PLoS ONE, 3*(2), e1679.

Chapter 20
Emotion Appears To Be Generated in REM Sleep in the Absence of Sensory Input. How Does This Finding Fit with Your Stimulus-Response Model of Emotion?

Joseph LeDoux

Emotions are generally thought of as states of the brain and body elicited by specific stimuli. Fear occurs when we encounter something threatening or dangerous. Rotten food elicits disgust. Sexual arousal occurs in the presence of appropriate partners. Stimuli that elicit emotions are said to have emotional potency or emotional competence. While it is easiest to study emotions elicited by stimuli in the external environment, emotions can also be elicited internally by thoughts and images, and presumably by such internal stimulation during dreams as well (LeDoux 1996).

As an example, consider the role of the amygdala in processing fear arousing stimuli. An external sound with innate emotional potency (a very loud sound or a sound associated with predatory species) or learned potency (a meaningless sound that has been associated with pain or other kinds of harm in the past) is processed by auditory regions in the thalamus and cortex that send connections to the lateral nucleus of the amygdala. By way of intra-amygdala circuits, the lateral amygdala then triggers responses of the central amygdala, which controls the expression of behavioral, autonomic, and endocrine responses that help the organism protect itself from the impending danger. Other sensory systems have similar connections with the lateral amygdala. Though over-simplified, this scheme illustrates the basis points of the circuitry. Yet, it only includes the inputs to the amygdala from sensory processing areas. In addition, the amygdala receives massive connections from a variety of cortical association areas in the temporal and frontal lobes, brain areas often associated with higher cognitive functions. Other important connections of the amygdala include the neuromodulatory systems in the brainstem that release monoamines throughout the forebrain and that play key roles in regulating sleep, wakefulness, and dreams (LeDoux 2007).

So there are ample ways in which internal stimuli in the form of images (in sensory processing areas) or thoughts (in higher cortical areas) could interact

J. LeDoux (✉)
Center for Neural Science, New York University, New York, NY, USA
e-mail: ledoux@cns.nyu.edu

N. Tranquillo (ed.), *Dream Consciousness*, Vienna Circle Institute Library 3,
DOI 10.1007/978-3-319-07296-8_20,

with generalized arousal in the brain (from monoamine activation) to trigger brain and bodily states typically associated with emotions. These states could then, in the context of a dream, add emotional coloration to the dream.

But another way to think about this is from the point of view of memory. Dreams are essentially explicit or declarative memories (or at least they are assessed as such memories). Amygdala activation has long been known to provide emotional coloration to declarative memories by interacting with the hippocampus directly and by way of amygdala activation of monoamines and hormonal responses (McGaugh 2003). By similar interactions during the dream, memories of dreams can be emotionally colored.

References

LeDoux, J. E. (1996). *The emotional brain*. New York: Simon and Schuster.
LeDoux, J. E. (2007). Amygdala. *Current Biology, 17*, R868–R874.
McGaugh, J. L. (2003). *Emotion and memory*. New York: Columbia U. Press.

Chapter 21
Do You Still Maintain That the Only Significant Difference Between Waking and REM Sleep-Dreaming Is Due to the Subtraction of Sensory Input in REM? What Is Your View of the Aminergic Demodulation Hypothesis That Derives from AIM?

Rodolfo R. Llinás

It is a pleasure to briefly address the very thoughtful lectures given by my esteemed colleague, Professor J. Allan Hobson. We have discussed, over the years, the issue of to what extent, from a brain function perspective, is dreaming different from the cognitive events we know as "being awake." To me, the posterior brain mechanisms involved in both dreaming and wakefulness are fundamentally similar. The important difference, from my viewpoint, resides in the fact that intentionality (a frontal lobe function) is not operant, i.e. we cannot guide our dreams in the same fashion we drive our thoughts and motricity when we are awake. Dreams simply happen to us in a willy-nilly fashion. This, by the way, is something we have determined experimentally in humans using magneto-encephalography.

The question then remains, given that this "hypo-frontality" may be the main difference between dreaming and wakefulness, what is the mechanism by which such difference is enacted? The explanation give by Professor Hobson is very attractive indeed. It makes sense that it should be the lower brainstem that dictates such a fundamental functional difference. Most particularly that it should be a chemical transmitter moiety, the aminergic system that normally enacts action, rather than contemplation, is a very provocative view. Thus a shutdown of such system could *de facto* result in imagery devoid of intention.

R.R. Llinás, M.D., Ph.D. (✉)
Department of Physiology and Neuroscience, NYU Medical School, New York, NY, USA
e-mail: rodolfo.llinas@med.nyu.edu

N. Tranquillo (ed.), *Dream Consciousness*, Vienna Circle Institute Library 3,
DOI 10.1007/978-3-319-07296-8_21, © Springer International Publishing Switzerland 2014

Chapter 22
What Is the Specific Significance of Dream Research for Philosophy of Mind?

Thomas Metzinger

Three examples: Altered states as contrast classes, self-model phase transitions in lucidity, and the devastating epistemological consequences of cognitive corruption.

Eminent philosophers like Daniel Dennett have convincingly demonstrated how far-reaching the conceptual and metatheoretical consequences of empirical research into dreaming are. From a philosopher's perspective, Allan Hobson's life work truly constitutes a major achievement that unveils such consequences and will continue to inspire generations of thinkers on mind, self, and consciousness – and I predict that it will do so in many different ways, and for decades to come. I, for one, have greatly profited from the seminal contributions made by empirical researchers like Hobson (who was always open to the philosophical dimension of his work), and empirically informed philosophers like Dennett (who has shown us how good philosophy of mind is not only about developing a conceptual commentary, but about shattering our intuitions). Here, I will select what, in my view, are the three most important issues connecting philosophy and dream research in this sense.

22.1 Altered States as Contrast Classes for Waking Consciousness

At the end of my popular book, *The Ego Tunnel*, I asked the following question: "*Can neurophenomenological refinement help us optimize critical scientific rationality? Could scientists be better scientists if they were well-traveled, say, if they learned to have lucid dreams?*" (Metzinger 2009, p. 239). Interestingly, one can answer the central methodological question about altered states of consciousness as

T. Metzinger (✉)
Philosophisches Seminar, Johannes Gutenberg-Universität Mainz, Mainz, Germany
e-mail: metzinge@uni-mainz.de

N. Tranquillo (ed.), *Dream Consciousness*, Vienna Circle Institute Library 3, DOI 10.1007/978-3-319-07296-8_22,

contrast classes for waking consciousness in two fundamentally different ways. First, from an objective, third-person point of view, the dream state provides the only naturally occurring *global* state of consciousness that we can use for a rigorous contrastive analysis, comparing relevant features of waking and dream consciousness. It is an easily replicated, but highly deviant model of phenomenal experience as a whole. Therefore, we can reliably use the dreaming brain as a global model system for waking consciousness, systematically comparing the statistical deep structure of autophenomenological reports produced by subjects in sleep laboratories with those of waking subjects. But we can also model the fine-grained functional profile of the neural correlates of dreaming and waking, describing their dynamical properties or the subtle information flow unfolding within them. And at least in principle, we can develop a more general mathematical model by "superimposing and subtracting," exposing the deeper functional structure connecting all phenomenal states in general. The activation-synthesis hypothesis (Hobson and McCarley 1977) and the AIM model (Hobson et al. 2000) show how this might be done by using a three-dimensional state-space to map phenomenological, functional, neurobiological and neurochemical differences between waking, REM sleep/dreaming and NREM sleep, and by suggesting how a number of other altered and pathological states of consciousness can be integrated into this model. A related and tremendously useful contribution is Hobson's suggestion of using dreaming as a contrast class – or indeed as a model – of delusional wake states (Hobson 1999). Philosophically, a deeply intriguing vision would be to extract the abstract commonalities marking out the causal structure that underlies the emergence of a phenomenal state space – for example, in all biological brains capable of waking and dreaming. A related point is Revonsuo's (2006) suggestion of using dreaming as a model system for the investigation of consciousness as such.

Then there is a second way of approaching the issue, namely by cultivating the scientist's mind itself or by exploring unusual contrastive conditions. Again, Hobson's work is a testament to his willingness to not only explore this territory, but show how first-person experience can be integrated into a broader theoretical framework, and he has repeatedly used his own dream reports to illustrate his claims about the formal features of dreaming (Hobson 2005, 2011). Though this point is often overlooked, philosophers can certainly profit from first-person familiarity with unusual contrastive conditions as well. Arguably, only those of us who have ever had a false awakening themselves can really profit from the deeply unsettling character of this experience. To wake up in a dream, only to later discover that really you are *still* dreaming, is not only a dramatic experience – it possesses great heuristic fecundity, because it forces you to think about things in a fundamentally new manner. It shatters many intuitions about the nature of reality and conscious experience you may have had before. For example, it throws an entirely new light – a *subjective* light – on problems related to the phenomenology of insight and certainty, and it suddenly makes it clear to the investigator how what today we still call "waking up" is something that comes in degrees and that can always and unexpectedly happen to you, at any point in time.

Another example is what one might call "mindful sleeping": If scientists were to systematically cultivate the art of going to sleep "meditatively", i.e., systematically practising a precise, gentle, and choiceless form of introspective awareness during the transition from waking to sleep, they could have the chance to consciously experience the onset of sleep paralysis, the transition from 2D hypnagogic imagery to full immersion in a dream scene (Nielsen 1991–1992, 1995; Windt 2014), or even spontaneously occurring out-of-body experiences themselves. They would now possess "first-person contrast conditions" to visual experience or the phenomenological profile of embodiment and agency in the wake state. And this would obviously change many of their implicit theoretical intuitions and strongly support the generation of new, testable hypotheses.

22.2 The Phenomenal Self in Transitions Between Ordinary and Lucid Dreaming

To me, it is even more fascinating to investigate the nature of the dream *self* and contrast it with the waking self as well as with different functional profiles *within* different kinds of dreaming. Philosophically, the change in self-consciousness during the transition from an ordinary dream to a lucid dream is most interesting (cf. Metzinger 2003, 2009, 2013b; Windt and Metzinger 2007 for details). What happens to the system's phenomenal self-model in the dream state, and what exactly happens when the system suddenly "realizes" that it is currently dreaming, i.e., when it suddenly begins to consciously experience itself *as* a dreamer? As it turns out, more than one kind of lucidity exists and can be conceptually and empirically distinguished (Windt and Metzinger 2007; Noreika et al. 2010). And again, Hobson and colleagues have contributed to this process by showing how the differences between wakefulness, lucid, and nonlucid dreams can be investigated and which different factors can be distinguished, both between and within lucid and nonlucid dreams – exactly the kind of progress I had secretly hoped for myself (Voss et al. 2009, 2013).

Not only are these discoveries of central importance for reframing philosophical concepts such as "mental autonomy" (Are you, the subject, really in control of your own mind? To *what degree* are you an autonomous cognitive agent?) or "epistemic transparency" (When exactly do you know *that you know* – in particular, when the question is applied to the special case of knowing your *own* mind?). For example, I would claim that the degree of mental autonomy exhibited by a given system is mainly a function first, of the number of distal goal representations it is able to sustain (which includes remembering past goal states and reaching out into the future) plus, second, the capacity for self-control via second-order mental action, e.g., the capacity for intentional inhibition and impulse control (cf. Metzinger 2013a, and especially 2013c for details and an extended discussion). This argument relates directly to the capacity of "plot control" in lucid control dreams: Can "mental autonomy" in the dream state be operationalized according these two dimensions? For a philosopher, the reframing of such traditional concepts can also lead to new theoretical questions, such as "What degree of lucidity, exactly, characterizes ordinary waking life?," but also to the formulation of and demand for new empirical research projects. We need more rigorous, high-quality work on lucid dreaming! And Allan Hobson (e.g., 2010) has long worked to promote exactly this point.

Interestingly, lucidity also is a property the dreamer can lose, repeatedly and unpredictably, but also *partially* – the transition process from ordinary to lucid dreaming can be investigated in two directions at once, and in a more fine-grained manner. Philosopher Jennifer Windt (2014) has pointed out how lucid lapses would offer a particularly interesting contrast condition for waking delusions. For instance, it seems that lucid dreamers, though realizing that they are dreaming, are still prone to believe that other dream characters are real (cf. Levitan 1994) – an important point, hitherto undiscovered in its theoretical potential. This might extend Hobson's familiar claim that dreaming is a model of psychosis (Hobson 1999; but see Windt and Noreika 2011) to include the contrast between *lucid* dreams and delusional wake states. Of even greater relevance is the recent discovery that phenomenal embodiment, or the full "lived body," is not necessary for spatial relatedness, and that there actually can be a sense in which spatial self-location is independent of bodily experience (though it will still involve the experience of being a minimally extended point; cf. Windt 2010). Here, we have a second example of how careful philosophical analysis of dreaming contributes to a concrete research project, namely of isolating the simplest form of self-consciousness and of specifying the necessary and/or sufficient conditions for minimal phenomenal selfhood (MPS; see Blanke and Metzinger 2009; Metzinger 2013b). In this way, dreaming can serve as a contrast condition for analyzing the structure of phenomenal experience independently of the physical body, but also of *phenomenal* embodiment. Of course, the same applies for different kinds of lucidity.

22.3 Cognitive Corruption

What is the relationship between hallucinating and thinking (Fosse et al. 2001)? And what distinguishes thinking during lucid and nonlucid dreams from each other, as well as from waking thought (Kahn and Hobson 2005)? A centrally important general discovery is that our own rationality is only given to us via the phenomenal self-model, via subjective experience – and that therefore we can never be sure to really *have* the cognitive capacities we subjectively experience ourselves as currently possessing (Metzinger 2013b, c). Put differently: You may have the robust experience of thinking logically and coherently, while this is actually not the case at all. You may be systematically unable to notice your own cognitive deficits, because they are simply not available to you on the level of transparent, conscious self-experience. You may hallucinate an intact memory without having one. You may also have the firm belief that you are awake, and that rational arguments and the reliability of your own cognitive capacities support your view while this is not at all the case. Of course, the same is true for lucidity. The belief that you have just discovered that you are currently dreaming could be just another illusory content on the level of the cognitive self-model, accompanied by an epistemically empty yet robust phenomenology of insight and certainty. It is a merit of Jennifer Windt to have isolated, analyzed and drawn attention to this point more clearly than anyone else. She writes:

> These instances of deception from corrupted cognition are disturbing from an epistemological point of view because they show that the mere availability of cognitive capacities says nothing about their reliability. In a dream, one can have the impression of engaging in rational thought or remembering something about one's waking life and be completely wrong. The phenomenology of knowing, thinking, and remembering seems particularly vulnerable to this type of corruption in the dream state. This type of dream deception, then, is not so much deception about the nature of the dream world as deception about the reliability of one's current cognitive abilities. It is epistemologically troubling because it brings the threat of deception even closer to home: whereas Cartesian dream deception has us deceived about the perceptual world and our bodies, deception from corrupted cognition has us deceived about our minds. Consequently, we can never be sure of being truly rational, at any given moment. (Windt 2014)

This last point is an excellent example of how modern dream research – as paradigmatically represented by the pioneering work of Allan Hobson – has highly relevant consequences for other disciplines, but also for the general image of man. Cartesian dream skepticism as discussed in classical philosophical epistemology only covered the *content* of dreams, but now we see how careful attention to the empirical facts forces philosophical epistemologists to accept a much more radical challenge: It is the *structure of cognition* itself, the coherence of the thinking self, which is in doubt at every single instant. And it is another point where dream research and philosophical analysis can complement each other, with the former contributing the empirical methods and isolating the underlying changes in brain activation patterns and neurochemical modulation (Hobson et al. 2000; Desseilles et al. 2011), and the latter contributing conceptual analysis and metatheoretical critique (Metzinger 2013b, c). The scientific investigation of corrupted cognition backfires into waking life, and as a matter of fact, it may even backfire into scientists academic activities in research and their epistemological self-conception. But it also shows us on a new level how fallible human beings actually are – whether waking or dreaming.

References

Blanke, O., & Metzinger, T. (2009). Full-body illusions and minimal phenomenal selfhood. *Trends in Cognitive Sciences, 13*(1), 7–13.

Desseilles, M., Dang-Vu, T. T., Sterpenich, V., & Schwartz, S. (2011). Cognitive and emotional processes during dreaming: A neuroimaging view. *Consciousness and Cognition, 20*(4), 998–1008.

Fosse, R., Stickgold, R. J., & Hobson, J. A. (2001). Reciprocal variation in thoughts and hallucinations. *Psychological Science, 12*(1), 30–36.

Hobson, J. A. (1999). *Dreaming as delirium. How the brain goes out of its mind.* Cambridge, MA: MIT Press.

Hobson, J. A. (2005). *13 dreams Freud never had: The new mind science.* New York: Pi Press.

Hobson, J. A. (2010). The neurobiology of consciousness: Lucid dreaming wakes up. *International Journal of Dream Research, 3*(1), 36–46. and the commentaries in the same volume.

Hobson, J. A. (2011). *Dream life: An experimental memoir.* Cambridge, MA: MIT Press.

Hobson, J. A., & McCarley, R. W. (1977). The brain as a dreamstate generator: An activation-synthesis hypothesis of the dream process. *American Journal of Psychiatry, 134*, 1335–1348.

Hobson, J. A., Pace-Schott, E. F., & Stickgold, R. (2000). Dreaming and the brain. Toward a cognitive neuroscience of conscious states. *Behavioral and Brain Sciences, 23*, 793–842.

Kahn, D., & Hobson, J. A. (2005). State-dependent thinking: A comparison of waking and dreaming thought. *Consciousness and Cognition, 14*(3), 429–438.

Levitan, L. (1994). A fool's guide to lucid dreaming. *NightLight,* 6. http://www.lucidity.com/NL62.FoolsGuide.html

Metzinger, T. (2003). *Being no one: The self-model theory of subjectivity.* Cambridge, MA: MIT Press.

Metzinger, T. (2009). *The ego tunnel: The science of the mind and the myth of the self.* New York: Basic Books.

Metzinger, T. (2013a). Two principles for robot ethics. In E. Hilgendorf & J.-P. Günther (Eds.), *Robotik und Gesetzgebung*. Baden-Baden: Nomos.

Metzinger, T. (2013b). Why are dreams interesting for philosophers? The example of minimal phenomenal selfhood, plus an agenda for future research. *Frontiers in Psychology, 4*, 746.

Metzinger, T. (2013c). The myth of cognitive agency: Subpersonal thinking as a cyclically recurring loss of mental autonomy. *Frontiers in Psychology, 4*, 931.

Nielsen, T. A. (1991–1992) A self-observational study of spontaneous hypnagogic imagery using the upright napping procedure. *Imagination, Cognition and Personality, 11*(4), 353–366.

Nielsen, T. A. (1995). Describing and modeling hypnagogic imagery using a systematic self observation procedure. *Dreaming, 5*(2), 75–94.

Noreika, V., Windt, J. M., Lenggenhager, B., & Karim, A. A. (2010). New perspectives for the study of lucid dreaming: From brain stimulation to philosophical theories of self-consciousness. Commentary on "The neurobiology of consciousness: Lucid dreaming wakes up" by J. Allan Hobson. *International Journal of Dream Research, 3*(1), 36–46.

Revonsuo, A. (2006). *Inner presence: Consciousness as a biological phenomenon.* Cambridge, MA: MIT Press.

Voss, U., Holzmann, R., Tuin, I., & Hobson, J. A. (2009). Lucid dreaming: A state of consciousness with features of both waking and non-lucid dreaming. *Sleep, 32*(9), 1191–1200.

Voss, U., Schermelleh-Engel, K., Windt, J. M., Frenzel, C., & Hobson, J. A. (2013). Measuring consciousness in dreams: The lucidity and consciousness in dreams scale. *Consciousness and Cognition, 22*, 8–21.

Windt, J. M. (2010). The immersive spatiotemporal hallucination model of dreaming. *Phenomenology and the Cognitive Sciences, 9*, 295–316.

Windt, J. M. (2014). *Dreaming: A conceptual framework for philosophy of mind and empirical research.* Cambridge, MA: MIT Press.

Windt, J. M., & Metzinger, T. (2007). The philosophy of dreaming and self-consciousness: What happens to the experiential subject during the dream state? In D. Barrett & P. McNamara (Eds.), *The new science of dreaming* (Cultural and theoretical perspectives, Vol. 3, pp. 193–248). Westport/London: Praeger Perspectives.

Windt, J. M., & Noreika, V. (2011). How to integrate dreaming into a general theory of consciousness – A critical review of existing positions and suggestions for future research. *Consciousness and Cognition, 20*(4), 1091–1107.

Chapter 23
How Does the Dream Consciousness/ Protoconsciousness Concept Resonate with Linguistic Ideas and the Hypothesis of a Universal Grammar?

Andrea Moro

We are living in an era when communication has never been easier. In the small world of the academy, we contact many more people every single day than any scholar of the pre-industrial era could have met in his entire life. The result, however, is that contacts tend to be reduced to electronic messages or, when time is generous, to video conferences on our laptops. Sadly, however, our correspondence is rarely guided by that sacred principle of chance which generates serendipity, one of the fundamental ingredients of science.

Meeting Allan Hobson inside the cloister of a wonderful monastery on the Venetian island of San Giorgio facing Piazza San Marco on a warm fall evening became for me not only a challenging opportunity for developing new ideas but a great personal pleasure. It all started when Allan asked me about some of my hypotheses. I was reporting the recent neuroimaging results that were obtained by making people learn "impossible languages" at an international conference (see Moro 2008). The discussion naturally turned into a dialogue where I learned new and exciting things about dreaming. Allan made me aware of one single crucial hypothesis (stemming from nineteenth century studies by Emil Kraepelin) that he developed within his own pioneering framework, namely that dreams are basically aphasic events. It was at this moment during the discussion when I perceived that a new world could be discovered by combining Allan's theories with the fundamentals of generative grammar.

You can easily imagine the first reaction of a linguist to such a hypothesis: we do have the impression of using the full-fledged linguistic competence during dreaming, so how can this hypothesis be true? Allan was patient and generous enough at that time to introduce me to the essential components of his theory of dreaming and I soon realized how really misleading our subjective impressions of dreaming may be on a more general ground, including our impression of consciousness. Indeed – to

A. Moro (✉)
Center for Neurolinguistics and Theoretical Syntax (NeTS), Institute of Advanced Study, IUSS, Pavia, Italy
e-mail: andrea.moro@iusspavia.it

N. Tranquillo (ed.), *Dream Consciousness*, Vienna Circle Institute Library 3, 167
DOI 10.1007/978-3-319-07296-8_23, © Springer International Publishing Switzerland 2014

put it in his own words – "we need language to give dream consciousness its narrative character and to make dream reporting possible" (Hobson 2002). But the language we find in dreaming is not the creative activity that we use in conscious life but rather an overimposed remembrance we add to the oneiric experience, similar to certain aspects of the visual remembrance we have in our dreams. Similarly, we can come to the same conclusion when it comes to logical thought. These questions in fact all point to a major crux that is, again in Allan's own words: "If logical thought is impaired, it may not be noticed – and explicitly acknowledged – because the very essence of self-reflective awareness is diminished" (Hobson et al. 2011).

This could have been enough to kill an evening conversation, but the smile on Allan's face and the nice Venetian atmosphere did the trick in a way that no text message nor video-conference could ever do. They fostered, or rather encouraged me, to keep talking. Ultimately, this discussion led me to a new question, a real gift in fact as all new questions, rather than answers, are. Quite possibly wrong, but – so it seems to me – well worth exploring. The following analysis is one linguist's reaction to Allan's seminal and thought provoking idea of dreaming as an aphasic (and illogical) activity related to a diminished state of consciousness.

The first step is to make clear that when we speak of language, we do intend a universe of phenomena, much as we do when we speak of the physical word. If we face this universe without wearing any thought-screen protector, we may just well stop there and stare off in mystical amazement. Linguists, instead, rather like to decompose such a vast marvel into several smaller pieces: they isolate the building blocks (lexicon), the rules of composition (syntax), or the possible sequences of sounds (phonology). Thus, when we speak of aphasic language, we should be very careful in specifying what type of structural entity any missing element belongs to. Another crucial fact that linguists know, ever since Galileo and Descartes in a metaphorical way and since Chomsky (1956) in a precise mathematical way, is that syntax is a unique trait of human languages. No animal can, by putting together the same elements in different sequences, form new and potentially infinite messages. For example, given three words like *Cain*, *Abel* and *killed* we can get both *Cain killed Abel* and *Abel killed Cain*, yielding two opposite meanings. In other words, all animals have dictionaries but humans have dictionaries made of words, whereas all other animals have dictionaries made of sentences.

Since the first pioneering works by Noam Chomsky in the 1950s, it is clear that syntax is governed by special mathematical structures: more specifically, it has a "recursive structure" in the sense specified in Chomsky (1956). The most advanced hypotheses on language structure consider recursion as the core aspect of syntax. In fact, recursion does not only characterize the compositional pattern of words. All rules of grammar (i.e., all principles ruling out ungrammatical combinations of words) can ultimately be traced back to the hierarchical structures created by recursive procedures rather than the linear sequences of words.

Among the several possible interpretations of the idea that dreaming is an aphasic activity, one can assume that dreaming lacks syntax in a sense that is coherent with the basic tenets of linguistics. This does not imply, of course, that the only way of interpreting "aphasia" in dreaming is to consider it as lacking syntax. Aphasia could well be referred to many other aspects related to language. Besides anomia – another typical possibility to interpret aphasia in dreaming – one can formulate many more complex issues related to language and dreaming. For example, one could construe meta-theoretical issues to explore such as determining if dreams can have language itself as an object of dreaming. One prototypical case study could be the case of poliglossy, as when one asks himself or herself in what language he or she is dreaming, or when one reflects on the appropriateness of a certain word or name of a person or a place, including the length or the category of a word. Nevertheless, the very hypothesis that dreaming is an aphasic activity is perfectly compatible with the idea that dreaming is an "asyntactic activity."

Can we find indirect evidence of this? This is what distinguishes – to me – a good question from a bad one: the very possibility to see if, independently from any answer, the question itself allows for a further question. And this seems to me to be the case. Let me just rephrase the conclusions that I reached after concluding our discussion that evening in Venice: first, dreams are aphasic activities; second, all and only human languages are characterized by syntax; third, syntax is characterized by recursion. The immediate consequence is that dreaming cannot be characterized by recursion. Again, how can this be indirectly tested?

It seems to me that there is at least one single striking implication in this conjecture: that dreaming is also an acalculic activity. Arythmetics is also characterized by recursive procedures, stemming from the definition of successor in Peano's Axioms (Odifreddi 1989). The sense upon which the very notion of numbers is formally built is essentially recursive in nature. The meeting between a leading psychiatrist, a neurologist, and a linguist has resulted in the birth of a new clue to explore and understand the nature of dreaming.

Is it true or at least worth investigating such a clue? I am certainly not able to provide a direct answer nor even a direct piece of evidence either way. But ever since that Venetian discussion (and many others, unfortunately via email), I have been recording my intense oneiric activity and interviewing reliable people on their own dreaming. Not in one single case did I get a response that the person was carrying out any calculation during dreaming. Of course, people can dream about single ciphers or more complex numbers – and thus make the Italian lottery flourish – or even dream about teaching mathematics, as it happened to some professor friends of mine. But they never had the impression of following a formal procedure for obtaining a new unknown result. The parallel with language is striking: we can dream of single words or of single numbers but we cannot compose them together with the same recursive procedures we exploit when we are conscious. Many more questions spring from this first conjecture. It could be very interesting to see if the neural circuits that correlate with syntax are active when dreaming since it has been proven that those circuits are sensitive to the presence/absence of recursive rules (see Moro 2008; Kandel et al. 2013 for a synthesis of the neurobiological foundations of language).

This short note reporting our first encounter ends here but it may well be the start of a new path to follow, one that stems from the intuitions and theories of Allan

Hobson and Noam Chomsky (Chomsky 2013) and one that offers us the much too infrequent sensation that good ideas can arise from personal and unprogrammed contact among people.

References

Chomsky, N. (1956). Three models for the description of grammar. *I.R.E. Transaction on Information Theory, 2*, 113–124.

Chomsky, N. (2013). Problems of projections. *Lingua, 130*, 33–49.

Hobson, J. A. (2002). *Dreaming. A very short introduction*. Oxford: Oxford University Press.

Hobson, J. A., Sangsanguan, S., Arantes, H., & Kahn, D. (2011). Dream logic: The inferential reasoning paradigm. *Dreaming, 21*(1), 1–15.

Kandel, E., Schwartz, J., Jessell, T., Siegelbaum, S., & Hudspeth, A. (2013). *Principles of neural science*. New York: McGrawHill Medical.

Moro, A. (2008). *The boundaries of Babel. The brain and the mystery of impossible languages*. Cambridge, MA: MIT Press.

Odifreddi, P. (1989). *Classical recursion theory*. Amsterdam: Elsevier.

Chapter 24
You Have Interpreted the PGO Waves of REM Sleep as Activation of the Startle Network of the Brain. What Is Your Theory of the Function of Off-Line Startle and What Impact, If Any, Does This Activation Have Upon Dreaming?

Adrian R. Morrison

24.1 The Meaning of PGO Waves

In 1976, Bob Bowker and I reported that waves seemingly identical to spontaneous PGO waves (usually called spikes then) could be elicited by sounds in REM sleep (Bowker and Morrison 1976) as well as NREM sleep. We proposed that ponto-geniculo-occipital (PGO) waves were probably evidence of the spontaneous excitation of the startle network. This conclusion stemmed from Dement's report (Ferguson et al. 1970) that cats under the influence of the anti-serotonergic drug, p-chlorophenylalanine (PCPA), startled every time a PGO wave occurred when they were "awake." We later changed from "startle" to "alerting" because we considered that the waves reflected the activation of a system responsible for detection of a stimulus, fundamental to the organization of the critical behavioral response of orienting, with startle being a less organized expression of a response to an unexpected stimulus. The focus on eye activity and dreams may have led other workers away from what is in our opinion their real significance: a sign of internal alerting in the brain (as distinct from behavioral *arousal*), a fundamental and "peculiar" aspect of REM sleep (Morrison and Bowker 1975).

Jouvet and Delorme (Jouvet and Delorme 1965) and later Henley and I (Henley and Morrison 1969; Henley and Morrison 1974) found that relatively small bilateral lesions in the rostral pontine tegmentum released cats from the grip of paralysis during REM sleep. Upon entering REM sleep, such cats will raise their heads with the third eyelids partially prolapsed across the eyes as in normal sleep but with the ears and vibrissae pricked forward as if the cat were alert and on the prowl. They are able to support themselves and pounce as if attacking imaginary prey. One might

A.R. Morrison, D.V.M., Ph.D. (✉)
Department of Animal Biology, Laboratory for the Study of the Brain in Sleep, School of Veterinary Medicine, University of Pennsylvania, Philadelphia, PA, USA
e-mail: armsleep@vet.upenn.edu

N. Tranquillo (ed.), *Dream Consciousness*, Vienna Circle Institute Library 3,
DOI 10.1007/978-3-319-07296-8_24,

conclude that their brains are activated in the same manner as in wakefulness but that they are "dreaming" or hallucinating, for the stimuli for these behaviors arise internally. Cats spend most of their waking hours always on the alert or stalking so that the behaviors we see in sleep must activate the same neural systems activated in wakefulness, not unlike the case for humans, although our lives and dreams are far more complicated.

At the cellular level, Steriade's group found in 1989 that PGO waves in the lateral geniculate body of cats resulted from nicotinic activation of the projecting neurons with a parallel muscarinic inhibition of perigeniculate cells stemming from activation by pontine peribrachial cholinergic neurons. They confirmed that auditory stimuli elicited PGO waves in the lateral geniculate body and concluded, in agreement with the earlier work, that "these signals are the central correlates of orienting reactions elicited by sensory stimuli during waking (the so-called eye movement potentials) and by internally generated drives during paradoxical sleep."

More recent work with tone-elicited waves in wakefulness in our laboratory has reinforced our original hypothesis. Using cats, we found that tones that elicited PGO waves (PGO_E) in the lateral geniculate body during REM sleep also elicited waves of comparable amplitude if the animal oriented to the tone source while awake. When the animal did not orient, the waves were of smaller amplitude even though each tone elicited a wave. We concluded "that the mere presence of PGO_E indicates that stimulus information has been registered, and that brain alerting mechanisms have been readied, though an overt behavioral response may not be forthcoming... Activation of central orienting mechanisms is reflected in larger amplitude $PGO_{E.}$ The presence of high-amplitude PGO_E accompanying orienting in waking, the high-amplitude PGO waves of REM, and the spontaneous 'orienting' of REM without atonia, support the idea that certain brain orienting mechanisms are spontaneously activated during REM. Indeed, the brain in REM and alert wakefulness is functioning in essentially similar ways." We reasoned that PGO_E of wakefulness signaled the enhancement of information transfer in the lateral geniculate body as revealed by Wolf Singer's group when they stimulated the reticular formation, an obvious benefit to visual orienting in cats (Rodriguez et al. 2004).

The importance of considering behavior of an animal when drawing conclusions about central events was brought home to us when considering the differences found between eye movement potentials (EMP) of wakefulness and PGO waves of REM sleep in their relation to eye movements. Earlier workers had observed that the cortical component of EMP was greatly diminished or absent in the dark. Reasoning that in the dark their cats had relaxed, we alerted them with various stimuli and observed that the EMP returned to their previous amplitudes.

I began by noting that suppression of serotonin resulted in the release of PGO waves. This, of course, also occurs as a cat begins to enter REM sleep, a state akin to continuous "orienting." Barry Jacobs and his colleagues found that a variety of physiological and environmental challenges, such as manipulations of glucose levels and blood pressure, a period of restraint, loud white noise, and a variety of others failed to affect firing of dorsal raphé serotonergic neurons; but presentation of a stimulus to which the awake cat oriented caused an immediate cessation of firing when the animal oriented (Jacobs et al. 1972; Heym at al. 1982). One might assume that had PGO_E been recorded they would have made their appearance as the serotonergic firing ceased.

References

Bowker, R. M., & Morrison, A. R. (1976). The startle reflex and PGO spikes. *Brain Research, 102* (1), 185–190.

Ferguson, J., Henriksen, S. J., Cohen, H., Mitchell, G., Barchas, J. D., & Dement, W. C. (1970). "Hypersexuality" and behavioral changes in cats caused by administration of p. chlorophenylalanine. *Science, 168*, 499–501.

Henley, K., & Morrison, A. R. (1969). Release of organized behavior during desynchronized sleep in cats with pontine lesions. *Psychophysiology, 6*, 245.

Henley, K., & Morrison, A. R. (1974). A re-evaluation of the effects of lesions of the pontine tegmentum and locus coeruleus on phenomena of paradoxical sleep in the cat. *Acta Neurobiologiae Experimentalis (Wars.), 34*(2), 215–232.

Heym, J., Trulson, M. E., & Jacobs, B. L. (1982). Raphé unit activity in freely moving cats: Effects of phasic auditory and visual stimuli. *Brain Research, 232*, 29–39.

Jacobs, B., Henriksen, S., & Dement, W. (1972). Neurochemical bases of the PGO wave. *Brain Research, 48*, 406–411.

Jouvet, M., & Delorme, F. (1965). Locus coeruleus et sommeil paradoxal. *Comptes Rendus Societe de Biologie, 159*, 895–899.

Morrison, A. R., & Bowker, R. M. (1975). The biological significance of PGO spikes in the sleeping cat. *Acta Neurobiologiae Experimentalis, 35*, 821–840.

Rodriguez, R., Kallenbach, U., Singer, W., & Munk, M. H. (2004). Short- and long-term effects of cholinergic modulation on gamma oscillations and response synchronization in the visual cortex. *Journal of Neuroscience, 24*(46), 10369–10378.

Chapter 25
What Is the Current Status of Your "Covert REM Process" Theory, Especially in the Light of the New Protoconsciousness Hypothesis?

Tore Nielsen

25.1 The Covert REM Sleep Processes Model of Dreaming and Protoconsciousness

The widespread use of standard polysomnographic (PSG) criteria for identifying sleep stages (Rechtschaffen and Kales 1968) has helped promote an artificially categorical view of sleep and the unproven notion that sleep stages are *mutually exclusive states separated by abrupt transitions*. Rather, evidence suggests that sleep stages consist of fluid, dissociable clusters of processes, with overlaps and admixtures being common under various circumstances (Nielsen 2000b; Himanen and Hasan 2000). This emerging, alternative view of sleep has important ramifications for theories of dreaming—and thus for theories of dream function like the protoconsciousness (ProtoCs) theory.

The covert-REM (C-REM) model of dream formation (refer to Fig. 25.1, Nielsen 2000b) applies the notion of stage admixtures to explain both why vivid, emotion-laden dreaming most typically occurs during REM sleep and why a less vivid form of dreaming also occurs, but with a lower frequency, during NREM sleep. The model proposes that some REM sleep processes responsible for dreaming also become active in a dissociated or covert manner during NREM sleep, especially immediately before or after REM sleep periods and at sleep onset. Others have suggested that such dissociations may also account for wake/dream admixtures (Ogilvie et al. 2003) and thus implicate covert REM sleep in some forms of waking consciousness. C-REM processes may be identified as partial PSG signs of REM sleep as these are reflected in standard measures of eye movements, muscle twitches, and EEG changes; but they may also appear as changes in a number of other measures that are highly characteristic of REM sleep, such as penile erections,

T. Nielsen (✉)
Département de Psychiatrie, Université de Montréal, Montréal, QC, Canada
e-mail: tore.nielsen@yahoo.com

N. Tranquillo (ed.), *Dream Consciousness*, Vienna Circle Institute Library 3,
DOI 10.1007/978-3-319-07296-8_25,

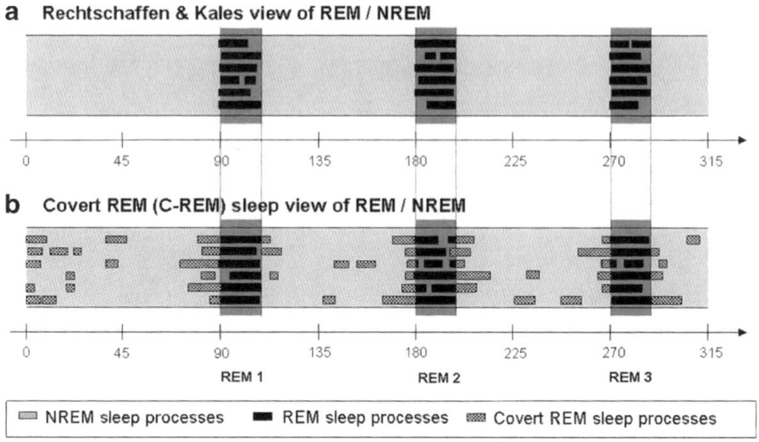

Fig. 25.1 (**a**) REM and NREM sleep are typically thought to be comprised of processes falling within the classical boundaries defined by standard criteria. (**b**) The C-REM model stipulates that they can also occur outside of these boundaries in normal sleep and, especially, during REM sleep deprivation

respiratory pauses, increases in HR variability, trains of EEG sawtooth waves, and so forth. In briefest outline, the C-REM model rests upon four premises:

1. REM and NREM sleep consist of organized, recurring clusters of physiological processes;
2. REM sleep mentation is produced by some subset of these processes (*isomorphism assumption*);
3. Some processes are dissociable from REM into NREM sleep (*dissociation assumption*);
4. Dissociated REM sleep processes can produce vivid mentation in NREM sleep.

Premise 1 is not typically disputed, however, it is often not appreciated that of the hundreds of physiological processes fluctuating during sleep, standard sleep staging is based upon only three easily recorded processes (EEG, EOG, EMG). Numerous other, less obvious, processes are not measured, indeed may not yet have been identified—and these may not adhere as clearly to the standard REM/NREM boundaries. Premise 2 is disputed by some (Pivik 1991) but there is evidence (Nielsen 2000b) for simple one-to-one isomorphisms between cognitive and physiological sleep variables, especially between dream and EEG measures (Germain et al. 1999) and dream emotion and autonomic measures. There is a near total lack of research examining one-to-many and many-to-many groupings of psychophysiological relationships. Accumulating evidence supporting premise 3 includes studies (Mahowald and Schenck 1992) of sleep/wake state dissociations in sleep disordered patients (e.g., narcolepsy, sleep paralysis) and healthy subjects. Among the latter, REM-related processes are regularly seen during NREM-to-REM transitional states, and include shifts in HR variability (Scholz et al. 1997), peripheral vasoconstriction (Lavie et al. 2000; Henane et al. 1977), and the fast EEG phases of the cycling alternating pattern or CAP (Terzano et al. 2005). Many other studies (Toth 1970) report partial signs of REM sleep, such as eye movements and EMG bursts, occurring during sleep onset (SO) NREM sleep (Nielsen 2000b). The high frequency of SO REM periods (SOREMPs) in healthy subjects may also be indicative of C-REM processes as these are more likely following increased REM propensity or sleep disruption (Takeuchi et al. 2002). While there is yet no systematic evidence on this point, it may be that C-REM processes are more prevalent among children and adolescents; SOREMPs are more prevalent among these groups (Carskadon et al. 1987, 1998) than among adults (Bishop et al. 1996), and dream-enacting behaviors—another form of REM state dissociation—are also very prevalent among the young (Fisher et al. 1989). Note that premise 3 is intended to include REM-related processes that are triggered prior to REM onset or may continue their activity after REM onset and whose dissociative nature is thus more of stage overlap than of 'fragmentation' from REM sleep. Premise 4 is supported by studies showing more frequent and vivid dream mentation from NREM sleep episodes that are sampled shortly after REM episodes (Nielsen 2000b).

Thus, the critical assumptions #3 and #4 of the C-REM model refer to physiological and cognitive evidence of state dissociations respectively. Apart from the pre-2000 literature supporting these two assumptions (see reviews: Nielsen 2000a, b), over the last decade several new supportive studies have appeared. *The authors of at least three independent replication studies all conclude that their findings support the C-REM model.* The following briefly summarizes this evidence.

Physiological Support

- Epochs of REM-like muscle atonia were found to occur very frequently in NREM sleep and to be increased after REM sleep deprivation, prompting the suggestion that they constitute the best available biomarker of REM sleep propensity (Werth et al. 2002; Tinguely et al. 2006).
- Parahippocampal EEG slow activity (1.5–3.0 Hz), a known REM sleep biomarker (Bódizs et al. 2001), is found regularly at SO (Bódizs et al. 2004). Alpha power decreases are also characteristic of both REM and SO (Bódizs et al. 2004).
- There is a quantifiable similarity between SO and whole night REM sleep in the EEG power spectrum (Bódizs et al. 2008).

Cognitive Support

- Two studies (Esposito et al. 2004; Germain and Nielsen 2001) show that recall of sleep mentation from SO, REM, or stage 2 NREM sleep is associated with a decrease in alpha EEG power.
- REM sleep deprivation increases the dreamlike quality of both later REM mentation and NREM mentation (Nielsen et al. 2005).
- Subjects administered an ultra short, 20 m/40 m, sleep-wake cycle for 72 h reported NREM dream recall whose circadian fluctuations almost exactly mirror ($r = 0.87$) circadian fluctuations in REM sleep propensity (Suzuki et al. 2004).

Further, delaying SO by 3 h increases NREM mentation vividness (Wamsley et al. 2004)—presumably because circadian-modulated REM propensity is elevated for those NREM sleep periods (Nielsen 2011).

In sum, the C-REM model of dreaming has garnered recent support from at least three independent laboratories for key physiological and cognitive assumptions. Note that one study using clomipramine to partially suppress REM sleep (Oudiette et al. 2012) found inconclusive evidence concerning the model.

Relationship to the ProtoCs theory. Because the C-REM model was formulated to account for methodological deficiencies in dreaming research, it speaks only indirectly to the ProtoCs theory. However, accumulating evidence does raise some possibilities for how the ProtoCs model may be tested.

Physiological basis. The C-REM model recognizes genetically determined biorhythmic influences on dream formation and is thus syntonic with the emphasis on ProtoCs as a causally critical biological process. In particular, the various biorhythmic influences on dream prevalence, frequency, and content (see review: Nielsen 2011), including ultradian, circadian, and sleep-dependent factors, are largely covert in that they operate across all sleep/wake stages. To illustrate, experimentally increased REM sleep propensity remains elevated, although not expressed, during all sleep/wake states; some evidence even suggests that elevated REM propensity may influence other states, i.e., by reducing the intensity of EEG slow-wave activity during NREM sleep (Beersma et al. 1990) and possibly reducing alpha power during wakefulness (Brunner et al. 1993). A similar type of covert process would appear to be needed to account for how the ProtoCs dream state influences the emergence of waking conscious states.

Developmental milestones. Hobson conceptualizes the ProtoCs state of dreaming as a primordial state of brain organization that serves as a building block for later consciousness. ProtoCs is proposed to develop into full-blown 'secondary Cs' as brain development proceeds *in utero* and through early life. We know little about the dreams of young children other than their structure appears to progress through stages that parallel the child's waking state cognitive capacities (Foulkes 1999). For example, between ages 6 and 7 years, dream imagery shifts from static to kinematic while between ages 7 and 8 years, the dream self-character role shifts from a passive-observer to an active-participant. If dreaming is indeed an essential precursor to mature consciousness, as the ProtoCs theory stipulates, then such changes in dreaming should regularly *precede* key milestones in the development of consciousness. If REM sleep stage dissociations are, in fact, more frequent among children, as indicated above, then these too may be implicated in the development of consciousness.

NREM mentation. Vivid dreaming and ProtoCs do not seem to be fundamental features of NREM mentation. A basic distinction between REM and NREM mentation pivots on the fact that REM dreaming is perceptually and emotionally vivid, two features that distinguish primary from secondary consciousness as well. A better understanding of what and how dreaming manifests in NREM sleep may help clarify if ProtoCs also depends upon NREM sleep and why it is more preferentially linked to REM sleep.

Learning/memory functions of sleep. In the last few decades, studies have shown that both REM and NREM sleep are implicated in learning and memory consolidation functions. It is increasingly likely that both sleep states work complementarily for maximal efficiency of sleep-related learning. The C-REM model finds an analogue in the 'sequential hypothesis' of learning which stipulates, for example, that avoidance learning in rats is associated with frequent 'transitional sleep,' or NREM sleep with REM partial signs (see review: Ambrosini and Giuditta 2001). It might be expected that ProtoCs, as a fundamental cognitive process (or the brain's 'most glorious achievement' according to Hobson), also depends upon such complex sleep stage progressions. Clearly, theories and evidence describing how consciousness is constituted from more basic physiological and cognitive processes are required.

References

Ambrosini, M. V., & Giuditta, A. (2001). Learning and sleep: The sequential hypothesis. *Sleep Medicine Reviews, 5*, 477–490.

Beersma, D. G. M., Dijk, D. J., Blok, C. G. H., & Everhardus, I. (1990). REM sleep deprivation during 5 hours leads to an immediate REM sleep rebound and to suppression of non-REM sleep intensity. *Electroencephalography and Clinical Neurophysiology, 76*, 114–122.

Bishop, C., Rosenthal, L., Helmus, T., Roehrs, T., & Roth, T. (1996). The frequency of multiple sleep onset REM periods among subjects with no excessive daytime sleepiness. *Sleep, 19*, 727–730.

Bódizs, R., Kantor, S., Szabo, G., Szucs, A., Eross, L., & Halasz, P. (2001). Rhythmic hippocampal slow oscillation characterizes REM sleep in humans. *Hippocampus, 11*, 747–753.

Bódizs, R., Sverteczki, M., Halász, P. (2004, October 5–9) *A parahippocampal electrocorticographic analysis of the transition from wakefulness to sleep in humans: Are REM features present?* In: 17th Congress of the European Sleep Research Society, Prague, Czech Republic.

Bódizs, R., Sverteczki, M., & Meszaros, E. (2008). Wakefulness-sleep transition: Emerging electroencephalographic similarities with the rapid eye movement phase. *Brain Research Bulletin, 76*, 85–89.

Brunner, D. P., Dijk, D. J., & Borbely, A. A. (1993). Repeated partial sleep deprivation progressively changes in EEG during sleep and wakefulness. *Sleep, 16*, 100–113.

Carskadon, M. A., Keenan, S., & Dement, W. C. (1987). Nighttime sleep and daytime sleep tendency in preadolescents. In C. Guilleminault (Ed.), *Sleep and its disorders in children* (pp. 43–52). New York: Raven Press.

Carskadon, M. A., Wolfson, A. R., Acebo, C., Tzischinsky, O., & Seifer, R. (1998). Adolescent sleep patterns, circadian timing, and sleepiness at a transition to early school days. *Sleep, 21*, 871–881.

Esposito, M. J., Nielsen, T. A., & Paquette, T. (2004). Reduced alpha power associated with the recall of mentation from Stage 2 and Stage REM sleep. *Psychophysiology, 41*, 288–297.

Fisher, B. E., Pauley, C., & McGuire, K. (1989). Children's sleep behavior scale: Normative data on 870 children in grades 1 to 6. *Perceptual and Motor Skills, 68*, 227–236.

Foulkes, D. (1999). *Children's dreaming and the development of consciousness*. Cambridge, MA: Harvard University Press.

Germain, A., & Nielsen, T. A. (2001). EEG power associated with early sleep onset images differing in sensory content. *Sleep Research Online, 4*, 83–90.

Germain, A., Nielsen, T. A., Khodaverdi, M., Bessette, P., Faucher, B., & Raymond, I. (1999). Fast frequency EEG correlates of dream recall from REM sleep. *Sleep, 22*(Suppl 1), S131–S132.

Henane, R., Buguet, A., Roussel, B., & Bittel, J. (1977). Variations in evaporation and body temperatures during sleep in man. *Journal of Applied Psychology, 42*, 50–55.

Himanen, S. L., & Hasan, J. (2000). Limitations of Rechtschaffen and Kales; review article. *Sleep Medicine Reviews, 4*, 149–167.

Lavie, P., Schnall, R. P., Sheffy, J., & Shlitner, A. (2000). Peripheral vasoconstriction during REM sleep detected by a new plethysmographic method. *Nature Medicine, 6*, 606.

Mahowald, M. W., & Schenck, C. H. (1992). Dissociated states of wakefulness and sleep. *Neurology, 42*(Suppl 6), 44–52.

Nielsen, T. A. (2000a). A review of mentation in REM and NREM sleep: 'covert' REM sleep as a possible reconciliation of two opposing models. *Behavioral and Brain Sciences, 23*, 851–866.

Nielsen, T. A. (2000b). Covert REM sleep effects on NREM mentation: Further methodological considerations and supporting evidence. *Behavioral and Brain Sciences, 23*, 1040–1057.

Nielsen, T. (2011). Ultradian, circadian, and sleep-dependent features of dreaming. In M. Kryger, T. Roth, & W. C. Dement (Eds.), *Principles and practice of sleep medicine* (5th ed., pp. 576–584). New York: Elsevier.

Nielsen, T. A., Stenstrom, P. M., Takeuchi, T., et al. (2005). Partial REM sleep deprivation increases the dream-like quality of mentation from REM sleep and sleep onset. *Sleep, 28*, 1083–1089.

Ogilvie, R. D., Takeuchi, T., & Murphy, T. I. (2003). Expanding Nielsen's covert REM model, questioning Solms's approach to dreaming and REM sleep, and reinterpreting the Vertes & Eastman view of REM sleep and memory. In E. F. Pace-Schott, M. Solms, M. Blagrove, & S. Harnad (Eds.), *Sleep and dreaming: Scientific advances and reconsiderations* (pp. 193–194). Cambridge: Cambridge University Press.

Oudiette, D., Dealberto, M. J., Uguccioni, G., et al. (2012). Dreaming without REM sleep. *Consciousness and Cognition, 21*, 1129–1140.

Pivik, T. (1991). Tonic states and phasic events in relation to sleep mentation. In S. Ellman & J. Antrobus (Eds.), *The mind in sleep. Psychology and psychophysiology* (2nd ed., pp. 214–248). New York: Wiley.

Rechtschaffen, A., & Kales, A. (1968). *A manual of standardized terminology, technique and scoring system for sleep stages of human subjects*. Bethesda: HEW Neurological Information Network.

Scholz, U. J., Bianchi, A. M., Cerutti, S., & Kubicki, S. (1997). Vegetative background of sleep—spectral analysis of the heart rate variability. *Physiology and Behavior, 62*, 1037–1043.

Suzuki, H., Uchiyama, M., Tagaya, H., et al. (2004). Dreaming during non-rapid eye movement sleep in the absence of prior rapid eye movement sleep. *Sleep, 27*, 1486–1490.

Takeuchi, T., Fukuda, K., & Murphy, T. I. (2002). Elicitation of sleep-onset REM periods in normal individuals using the sleep interruption technique (SIT). *Sleep Medicine, 3*, 479–488.

Terzano, M. G., Parrino, L., Smerieri, A., et al. (2005). CAP and arousals are involved in the homeostatic and ultradian sleep processes. *Journal of Sleep Research, 14*, 359–368.

Tinguely, G., Huber, R., Borbely, A. A., & Achermann, P. (2006). Non-rapid eye movement sleep with low muscle tone as a marker of rapid eye movement sleep regulation. *BMC Neuroscience, 7*, 2.

Toth, M. F. (1970). A new method for detecting eye movement in sleep. *Psychophysiology, 7*, 516–523.

Wamsley, E. J., Hirota, Y., Tucker, M. A., et al. (2004). Circadian influences on sleep mentation. *Sleep, 27*(Suppl), A65–A66.

Werth, E., Achermann, P., & Borbely, A. A. (2002). Selective REM sleep deprivation during daytime. II. Muscle atonia in non-REM sleep. *American Journal of Physiology Regulatory, Integrative and Comparative Physiology, 283*, R527–R532.

Chapter 26
How Does the Theory and Data Discussed in the Lectures Fit with Your Work on Dreams?

Valdas Noreika

The protoconsciousness theory of dreaming, recently proposed by J. Allan Hobson (2009), provides a timely synthesis of dream related findings that are otherwise scattered across various disciplines of scientific enquiry, and suggests interesting new ways for how to conceptualize and empirically investigate subjective experiences during sleep. More than that, far reaching but somewhat hidden and rather speculative implications of the protoconsciousness theory yield shocking insights into who we might be and, especially, into how we may have come into existence as conscious beings. Assuming these implications will receive empirical confirmation in the future, the protoconsciousness theory of dreaming may become an integral part of scientific understanding of the development of consciousness.

Converging empirical evidence suggests that the human foetus spends most of its time in sleep, and that at the age of 28–30 weeks of gestation most of the sleep time is spent in active REM-like periods (Graven and Browne 2006; Peirano et al. 2003). Assuming that conscious experience emerges during the last trimester of foetal development, the most likely behavioural state for this to happen appears to be REM sleep. Consequently, the first quale a foetus experiences turns out to be a somnic quale, and the state in which we come to exist as conscious beings – although literally in a minimal form – is sleep! Moreover, waking experiences may appear to be secondary to experiences during sleep, i.e. waking consciousness of a neonate may depend on the neuronal capacities that the brain has developed for internal experiences during sleep by generating and maintaining sensory circuits. If so, then consciousness during sleep is no longer a reflection of waking life; instead experiences during sleep, at least during ontogenetic development, are prior to experiences that we have during wakefulness. Such a shift in emphasis may also

V. Noreika (✉)
Department of Psychology, Centre for Cognitive Neuroscience, University of Turku, Turku, Finland

Department of Psychiatry, Oxford Centre for Human Brain Activity (OHBA), Warneford Hospital, University of Oxford, Oxford, UK
e-mail: valnor@utu.fi

N. Tranquillo (ed.), *Dream Consciousness*, Vienna Circle Institute Library 3,
DOI 10.1007/978-3-319-07296-8_26,

shift the understanding of the origins of mental life – we are no longer primarily diurnal beings, we are primarily somniacs! No doubt our first foetal experiences during sleep were much simpler than our adult dreams, as they probably consisted of basic sensory impressions. Nevertheless, internally generated subjective experiences during REM sleep may have formed the primary virtual model of the world, which later became continuously updated during increasing periods of active wakefulness.

Interestingly, content analysis studies of adult dream reports, summarized by J. Allan Hobson in the target essays, show that our somnic mental life is largely psychotic. In dreams we typically hallucinate, confabulate and are deluded to such an extent that passing of a simple mental state exam would not be possible. This is not surprising given the active mode of REM sleep generating mechanisms: the sleeping brain does not simply reactivate waking memories, but synthesizes them into a story, enforcing new associations between originally separate elements and filling in the missing gaps. As a result of such active processes, dream narratives tend to be as bizarre and unpredictable as are the waking states of psychotic patients (Noreika et al. 2010; Scarone et al. 2008).

When these two premises are coupled together, i.e. that conscious existence originates in sleep, and that mental states during sleep are largely psychotic, a surprising conclusion is reached: the primary form of conscious experience during foetal and neonatal development might be a somnic psychosis. That is, our waking consciousness not only stems from its somnic protoform, but this protoform is entirely hallucinatory and perceptually psychotic. Such speculations, derived from the protoconsciousness theory of dreaming, may fail empirical testing or may even prove to be too complicated to test experimentally, e.g. psychosis model of dreaming has a number of theoretical and practical restrictions (Windt and Noreika 2011). Nevertheless, the implications and predictions of the protoconsciousness theory may consolidate and guide empirical studies of consciousness in the future. If the emergence and development of sensory circuitries during foetal sleep leads to the emergence of subjective experiences, then a search for the neural correlates of consciousness may need to shift to the unexpected domain of foetal sleep. Finally, the protoconsciousness theory of dreaming may increase our empathy with mentally disturbed patients: we not only experience psychosis each night when we fall asleep, but possibly the simplest form of such somnic psychosis was the original conscious existence from which we grew up and simply forgot.

Acknowledgments I wish to thank Katja Valli and Jennifer M. Windt for many exciting discussions on dreaming and consciousness that shaped my own views, and the National Doctoral Programme of Psychology and the Academy of Finland for funding my research.

References

Graven, S. N., & Browne, J. V. (2006). Sleep and brain development: The critical role of sleep in fetal and early neonatal brain development. *Newborn and Infant Nursing Reviews, 8*(4), 173–179.

Hobson, J. A. (2009). REM sleep and dreaming: Towards a theory of protoconsciousness. *Nature Reviews Neuroscience, 10*(11), 803–813.

Noreika, V., Valli, K., Markkula, J., Seppälä, K., & Revonsuo, A. (2010). Dream bizarreness and waking thought in schizophrenia. *Psychiatry Research, 178*(3), 562–564.

Peirano, P., Algarín, C., & Uauy, R. (2003). Sleep-wake states and their regulatory mechanisms throughout early human development. *The Journal of Pediatrics, 143*(4 Suppl.), 70–79.

Scarone, S., Manzone, M. L., Gambini, O., Kantzas, I., Limosani, I., D'Agostino, A., & Hobson, J. A. (2008). The dream as a model for psychosis: An experimental approach using bizarreness as a cognitive marker. *Schizophrenia Bulletin, 34*(3), 515–522.

Windt, J. M., & Noreika, V. (2011). How to integrate dreaming into a general theory of consciousness – A critical review of existing positions and suggestions for future research. *Consciousness and Cognition, 20*(4), 1091–1107.

Chapter 27
How Does Protoconsciousness Theory Mesh with Your Model of Dream Emotion?

Edward F. Pace-Schott

The role of sleep and dreaming in maintaining emotional stability represents a very tangible and practical example of protoconsciousness as a mental state that supports the proper functioning of normal waking consciousness. Normal sleep has been shown to promote basic mammalian mechanisms of emotion regulation such as habituation, extinction, and physiological homeostasis (Pace-Schott et al. 2009a, b; McEwen 2006). Sleep deprivation experiments suggest that sleep is also essential to cognitively based emotion regulatory functions such as accurate identification of facial emotion (van der Helm et al. 2010). Dreaming has been widely hypothesized to take part in this emotion regulatory process. For example, Rosalind Cartwright has suggested that negative affect is progressively ameliorated across dreams elicited from successive REM periods of a night in mildly depressed college students (Cartwright et al. 1998a). Similarly, she has linked a pattern of progression from negative early dreams to positive late dreams across the night with remission at 1 year in persons meeting Beck Depression Inventory criteria for depression (Cartwright et al. 1998b). The pattern of brain activation across sleep stages revealed by PET studies, which show global de-activation in NREM followed by selective re-activation of limbic structures that include core elements of the brain's fear and reward processing networks, suggest that both positive and negative emotional extremes could be moderated during REM and that REM sleep dreaming may reflect a subjective experience of this process (Pace-Schott 2010). Indeed, Tore Nielsen and Ross Levin have suggested that these REM-activated limbic structures regulate emotion during REM sleep via extinction processes, and that, in PTSD, this process is disrupted resulting in both nightmares and impaired daytime emotion regulation (Levin and Nielsen 2007). Therefore, functionality in terms of emotional homeostasis has been attributed not only to the selectively activated physiology of REM itself but also to its subjective manifestation, REM sleep dreaming. Protoconsciousness theory posits "A primordial state of brain organization that is a building

E.F. Pace-Schott (✉)
Department of Psychology, University of Massachusetts, Amherst, MA, USA
e-mail: EPACE-SCHOTT@mgh.harvard.edu

N. Tranquillo (ed.), *Dream Consciousness*, Vienna Circle Institute Library 3,
DOI 10.1007/978-3-319-07296-8_27,

block for consciousness" (Hobson 2009). Hobson (2009) suggests that this primordial state of consciousness is prominent prenatally and in infancy when it supports the developing "secondary consciousness" of later childhood and adulthood. Hobson posits further that protoconsciousness then continues throughout life, especially during REM sleep dreaming, functioning in support of waking consciousness. If consciousness can be profitably described and compared between brain states in terms of its component formal domains, as suggested in Hobson's first lecture of the current series, then certainly the emotional domain is one in which support of waking function is ongoing and essential given the lifelong nature of stressors and other challenges to proper functioning of the emotional domain. And, as most clearly seen during acute stress or in the disorders of emotion (affective and anxiety), disregulation in the emotional domain has innumerable knock-on effects on all other realms of adult waking secondary consciousness impacting higher cognitive functions such as selective attention, ability to reason, and ability to plan prospectively. Therefore, the nightly support of waking consciousness, whether as a function of a protoconscious REM state or the physiological processes of sleep itself, represents an undeniable and essential function of sleep.

References

Cartwright, R., Luten, A., Young, M., Mercer, P., & Bears, M. (1998a). Role of REM sleep and dream affect in overnight mood regulation: A study of normal volunteers. *Psychiatry Research, 81*, 1–8.

Cartwright, R., Young, M. A., Mercer, P., & Bears, M. (1998b). Role of REM sleep and dream variables in the prediction of remission from depression. *Psychiatry Research, 80*, 249–255.

Hobson, J. A. (2009). REM sleep and dreaming: Towards a theory of protoconsciousness. *Nature Reviews Neuroscience, 10*, 803–813.

Levin, R., & Nielsen, T. A. (2007). Disturbed dreaming, posttraumatic stress disorder, and affect distress: A review and neurocognitive model. *Psychological Bulletin, 133*, 482–528.

Mcewen, B. S. (2006). Sleep deprivation as a neurobiologic and physiologic stressor: Allostasis and allostatic load. *Metabolism, 55*(Suppl 2), S20–S23.

Pace-Schott, E. F. (2010). The neurobiology of dreaming. In M. H. Kryger, T. Roth, & W. C. Dement (Eds.), *Principles and practice of sleep medicine* (5th ed.). Philadelphia: Elsevier.

Pace-Schott, E. F., Marcello, M., Shepherd, E., Propper, R. E., & Stickgold, R. (2009a). Effect of napping on inter-session habituation to emotional stimuli-preliminary psychophysiological and subjective results. *Sleep, 32*(Supplement), A419.

Pace-Schott, E. F., Milad, M. R., Orr, S. P., Rauch, S. L., Stickgold, R., & Pitman, R. K. (2009b). Sleep promotes generalization of extinction of conditioned fear. *Sleep, 32*, 19–26.

Van Der Helm, E., Gujar, N., & Walker, M. P. (2010). Sleep deprivation impairs the accurate recognition of human emotions. *Sleep, 33*, 335–342.

Chapter 28
Do You Suppose That, in Addition to the Sensorimotor Isolation of REM, There Is Impairment of Intrinsic Attentional Processes That We Experience as an Inability to Observe and Think in Our Dreams?

Michael Posner

I answer this question in the spirit of the general Hobson theory which seeks a mechanistic answer to Dream Consciousness. The problem of consciousness involves two aspects for which attention is central: awareness and control. Clearly we are aware in the dream state, but for most people the ability to be in control of the dream content is lost.

In recent years, the mechanisms of cognitive and emotional control have been studied in some detail within the field of attention. Imaging studies of executive attention indicate a brain network that includes the anterior cingulate gyrus, anterior insula and the underlying striatum (Petersen and Posner 2012; Posner and Rothbart 2007). It seems likely that this control network receives input from a goal tree that involves some of the prefrontal cortex. The goal tree represents the momentary list of goal states that provides coherence to our thoughts (Carbonell 1981; Duncan et al. 1996). In REM sleep, the goal tree and mechanisms of control are clearly impoverished.

Surprisingly, awareness of the external world and awareness of internal thoughts appear to be carried out by separate brain networks (Posner and Rothbart 2007). Orienting to sensory stimuli involves a specific cortical network that includes the inferior and superior parietal lobe and frontal eye fields. Awareness of internal thoughts and feelings involve the anterior cingulate and other areas relate to executive attention.

Early in life, cognitive control appears to depend more on an orienting network (Posner et al. 2012). This network underlies the young child's reliance on external input to provide control. In the absence of the more executive network during REM, it seems likely that the orienting network may be the main method of control. Of course external input is minimal during sleep, but it can either interrupt REM or be incorporated into the dream state.

M. Posner (✉)
Institute of Cognitive and Decision Sciences, University of Oregon, Eugene, OR, USA
e-mail: mposner@uoregon.edu

N. Tranquillo (ed.), *Dream Consciousness*, Vienna Circle Institute Library 3,
DOI 10.1007/978-3-319-07296-8_28,

It is known that training in lucid dreaming can improve the ability of the person to control the dream (Gackenbach and LaBerge 1988). It might be speculated the training somehow allows more normal use of the goal tree to influence the mechanisms of control.

Thus the REM state appears to dissociate two functions of attention awareness and control. Control is generally impaired although training may reduce its impairment. Awareness is unimpaired although its direction is to internal not external information.

References

Carbonell, J. G. (1981). *Subjective understanding, computer models of belief systems*. Ann Arbor: UMI Research Press.

Duncan, J., Emslie, H., Williams, P., Johnson, R., & Freer, C. (1996). Intelligence and the frontal lobe: The organization of goal-directed behavior. *Cognitive Psychology, 30*(3), 257–303.

Gackenbach, J., & LaBerge, S. (1988). *Conscious mind, sleeping brain: New perspectives on lucid dreaming*. New York: Plenum Press.

Petersen, S. E., & Posner, M. I. (2012). The attention system of the human brain: 20 years after. *Annual Review of Neuroscience, 35*, 71–89.

Posner, M. I., & Rothbart, M. K. (2007). Research on attention networks as a model for the integration of psychological science. *Annual Review of Psychology, 58*, 1–23.

Posner, M. I., Rothbart, M. K., Sheese, B. E., & Voelker, P. (2012). Control networks and neuromodulators of early development. *Developmental Psychology, 48*(3), 827–835.

Chapter 29
Does Your "Single-Minded" Characterization of Dreaming Now Find Confirmation and Explanation in the Lucid Dreaming Data?

Allan Rechtschaffen

The descriptions by Dr. Hobson and others of lucid dreaming (states in which the dreamer has a reflective awareness that he is dreaming) demonstrate nicely that what we have called the "the single-mindedness" (absence of that reflective awareness) in most dreams is not an inevitable attribute of dreaming *per se* (Rechtschaffen 1997). Rather, by techniques described by Dr. Hobson and others, lucidity can be induced in association with dreams that would otherwise be single-minded. In turn, this means that lucidity must originate from specific brain processes. Early imaging studies showed low levels of frontal lobe neural activity during REM sleep, suggesting that the mechanism for lucidity might be at least partly localized to that area. This hypothesis is bolstered by reports of dream-like mentation during wakefulness in patients with frontal lobe injury.

These observations further suggest that frontal lobe processes might be important for the maintenance of lucidity during normal wakefulness. Direct evidence for an association between lucidity and frontal lobe activity comes from the study of Voss et al. (2009), which showed an increase in high frequency frontal region EEG activity during lucid dreams. Altogether, the single-mindedness/lucidity/frontal lobe story is illustrative of a major theme of "The Essay" – the contribution of empirical sleep research to the elucidation of mind-body issues.

References

Rechtschaffen, A. (1997). Postscript to the single-mindedness and isolation of dreams. In M. S. Myslobodsky (Ed.), *The mythomanias: The nature of deception and self-deception* (pp. 203–223). Mahwah: Lawrence Erlbaum Associates.

Voss, U., Holzmann, R., Tuin, I., & Hobson, J. A. (2009). Lucid dreaming: A state of consciousness with features of both waking and non-lucid dreaming. *Sleep, 32*(9), 1191–1200.

A. Rechtschaffen (✉)
Departments of Psychiatry and Psychology, College University of Chicago, Chicago, IL, USA
e-mail: alrecht@sbcglobal.net

N. Tranquillo (ed.), *Dream Consciousness*, Vienna Circle Institute Library 3,
DOI 10.1007/978-3-319-07296-8_29, © Springer International Publishing Switzerland 2014

Chapter 30
The Idea, Championed by Your Group, That Dreaming Functions as Threat Avoidance Would Seem to Be Quite Compatible with Protoconsciousness Theory. But Is It Really Dreaming (as Against REM) That Performs That Function? In Other Words, Are You a Dualist or a Neutral Monist?

Antti Revonsuo

30.1 Introduction

"Protoconsciousness" is the central concept in Hobson's (2009) new theory. In this brief commentary, I will focus on two aspects of the protoconsciousness theory. First, what precisely *is* protoconsciousness? Is it a phenomenon of *consciousness*, although more "primitive" in form than our waking consciousness? Or is it an *unconscious*, purely neural process? This central question remains somewhat unclear in the theory. Second, what does the protoconsciousness theory say about the function of dreaming?

To answer both of these questions, I will try to relate the protoconsciousness idea to the concepts and definitions of consciousness and dreaming that I have myself systematically developed in my earlier publications (Revonsuo 1995, 2000a, b, 2006, 2010).

30.2 Is Protoconsciousness Conscious or Unconscious?

Protoconsciousness is defined as "A primordial state of brain organization that is a building block of consciousness" (Hobson 2009, p. 808). Protoconsciousness develops in REM sleep. Dreamless, "protoconscious REM sleep could. . .provide a virtual world model, complete with an emergent imaginary agent. . .that moves

A. Revonsuo (✉)
School of Humanities and Informatics, University of Skövde, Skövde, Sweden

Department of Psychology, University of Turku, Turku, Finland
e-mail: revonsuo@utu.fi

N. Tranquillo (ed.), *Dream Consciousness*, Vienna Circle Institute Library 3,
DOI 10.1007/978-3-319-07296-8_30, © Springer International Publishing Switzerland 2014

through fictive space...and experiences strong emotion as it does so. This state is not at first associated with awareness" (Hobson 2009, p. 808).

This description suggests that protoconsciousness first appears at a level of brain organization that does not involve subjective phenomenality (phenomenal consciousness or phenomenal dreaming) at all, but is completely unconscious. Thus, protoconsciousness first appears at the purely neurophysiological levels of organization, not at the phenomenal level of organization.

It is difficult to see how this could be the case because, on the other hand, Hobson says that protoconsciousness is a virtual world model with fictive space, an agent moving through that space and experiencing strong emotion (Hobson 2009, p. 808). The notion of "virtual world" or "virtual reality" refers to the *experience* of a subject being immersed into a space. The agent, its movements, and the emotions experienced by that agent are arguably constituted by subjective experience. And, as is widely agreed, the notion of "phenomenal consciousness" refers to the presence or occurrence of subjective experiences. Thus, the problem is that Hobson seems to claim that protoconsciousness is *not* conscious, but at the same time he describes the features of protoconsciousness with terminology that necessarily implies subjective, qualitative phenomenal experiences, or phenomenal consciousness.

He furthermore says that "...only during childhood do we become able to integrate this experience and become aware of it. The consciousness that we come to enjoy in our dreams is predominantly primary..." (Hobson 2009, p. 808). This suggests that Hobson is perhaps committed to a Higher-Order Thought (HOT) theory of consciousness, according to which we have to become *aware of* our experiences before they become conscious. Another dream researcher who explicitly defended HOT theory and applied it to dreaming is David Foulkes (1999). I myself do not support this kind of interpretation of either dreaming or consciousness. HOT theories lead to what I consider a contradiction in terms: qualia without consciousness, or unconscious experiences. In my conceptual framework, this leads to the logically incoherent notion of nonphenomenal phenomenality.

Thus, I think that the ontological status of protoconsciousness needs to be clarified (is it conscious or nonconscious?; phenomenal or nonphenomenal?), as well as the philosophical theory of consciousness that is being used to define consciousness. Otherwise it is not easy to understand what exactly the theory says about the fundamental nature of dreaming and consciousness.

30.3 Dreaming as a Virtual Reality or World Simulation

Whatever the precise ontological status of protoconsciousness may be, I fully endorse the idea that dreaming is a kind of virtual reality. I have put forward this characterization of dreaming as a virtual reality in my earlier theoretical work on the definition of dreaming (Revonsuo 1995). More recently, I have also used the concept of "world-simulation" to characterize dreaming and to separate genuine dreaming from mere thought-like sleep mentation (Revonsuo 2006, 2010).

Dreaming is a simulated phenomenal world. A related definition of dreaming as "immersive spatiotemporal hallucination" has recently been put forward by Windt (2010) in a paper that gives a detailed conceptual analysis and clarification to ideas behind the virtual reality model of dreaming. Thus, Hobson's protoconsciousness theory is well in harmony with other current ideas that relate dreaming to virtual reality.

Hobson's REM-sleep-dream protoconsciousness hypothesis suggests that the development and maintenance of waking consciousness depends on brain activation during sleep. From this flows a new theory of consciousness that suggests that the brain states underlying waking and dreaming cooperate and that their functional interplay is crucial for the optimal functioning of both (Hobson 2009, p. 811).

Hobson's idea about the function of protoconsciousness and dreaming appears to be mostly developmental and happens at a very general level. During REM sleep, the brain simulates information about time and space, and uses its built-in predictions to do that. Those predictions are then adjusted by the real stimulation that is received during wakefulness. Simulation of time, space, agency, movement, and emotion, appears to be the primary function of Hobson's virtual reality simulator.

I have put forward a much more specific hypothesis concerning the simulation function that might be at play during dreaming: the Threat Simulation Theory (TST) of the function of dreaming (Revonsuo 2000a, b). According to this theory, the dreaming brain selects among emotionally negatively charged memories when it puts together a dream simulation. In our dreams, negative contents are much more likely to appear than positive ones. Dreams and nightmares simulate various dangerous and threatening events and force us to rehearse the brain mechanisms of the perceptual and motor skills that are needed in efficient threat recognition and avoidance. (The characterization of TST given here is necessarily extremely brief, for a detailed description of the theory, its predictions, and the evidence so far, see Revonsuo 2000a, b and the review in Valli and Revonsuo 2009).

The threat simulation idea is consistent with the general spirit of the protoconsciousness theory: the function(s) of dreaming are based on the simulations of real life events, and their functions are evolutionarily ancient. However, the TST issues much more precise hypotheses as to what kind of contents we should observe in dreams and under which circumstances. The theory predicts that certain types of content under certain types of conditions are frequently simulated because they are functionally valuable for the organism's survival (or were so during the evolutionary history of the species).

TST does not rule out that dreaming may have other, more general simulation functions than the simulation of threatening situations. TST merely states that threatening situations have had a very high survival value in our evolutionary past and therefore they still have a strong and predictable representation among our various dream simulations.

Hobson's functional simulation hypothesis is thus more at the level of the general form of dreaming as a simulation, whereas TST is at the level of specific dream contents and specific cognitive-behavioral survival skills. TST explicitly claims that dreaming about certain types of contents has an evolutionarily ancient survival function. TST also states that it is the phenomenal dreaming itself, the dream as an experience (rather than the lower-level neurophysiological mechanisms alone), that has this function (or at least that phenomenal dreaming is necessary for the function to be carried out, Revonsuo 2000a). Thus, the dream as experienced is not epiphenomenal.

There are both similarities and differences between Hobson's and my own theories, but they are consistent with each other. TST issues rather specific predictions as to what we should and should not observe in dream content. It is unclear to me how exactly we could test the protoconsciousness theory, and what its specific predictions are. Thus it would be valuable for the future of the protoconsciousness theory to formulate some empirically testable predictions that follow from it.

References

Foulkes, D. (1999). *Children's dreaming and the development of consciousness*. Cambridge, MA: Harvard University Press.

Hobson, J. A. (2009). REM sleep and dreaming: Towards a theory of protoconsciousness. *Nature Reviews Neuroscience, 10*, 803–813.

Revonsuo, A. (1995). Consciousness, dreams, and virtual realities. *Philosophical Psychology, 8*, 35–58.

Revonsuo, A. (2000a). Did ancestral humans dream for their lives? *Behavioral and Brain Sciences, 23*(6), 1063–1082.

Revonsuo, A. (2000b). The reinterpretation of dreams: An evolutionary hypothesis of the function of dreaming. *Behavioral and Brain Sciences, 23*(6), 877–901.

Revonsuo, A. (2006). *Inner presence. Consciousness as a biological phenomenon*. Cambridge, MA: MIT Press.

Revonsuo, A. (2010). *Consciousness, the science of subjectivity*. Hove: Psychology Press.

Valli, K., & Revonsuo, A. (2009). The threat simulation theory in the light of recent empirical evidence—a review. *The American Journal of Psychology, 122*, 17–38.

Windt, J. M. (2010). The immersive spatiotemporal hallucination model of dreaming. *Phenomenology and Cognitive Science, 9*, 295–316.

Chapter 31
What Are the Clinical Implications of Protoconsciousness Theory for the Conceptualization of Those Psychiatric Disorders Commonly Referred to as Mental Illnesses?

Silvio Scarone and Armando D'Agostino

J. Allan Hobson proposes that human high-order awareness, a type of meta-awareness, arose from a basic and low-order awareness ("protoconsciousness") common to animals and humans that is primarily emotional and perceptual. Hobson also suggests that the latter is phenomenally fully expressed in the dreams of both animals and humans, whereas the former is only expressed in humans during wakefulness and is intimately related to the development of language. According to this theory, the awareness of one's self and all related aspects of metacognition depend on the flowing of time and space related personal experiences which are stored in memory circuits after some mental elaboration mediated by linguistic abilities. In his model, the mutual relationships and influences between these two levels of human awareness must be physiologically regulated to avoid psychosis. In Hobson's view, dreaming is akin to an organic psychosis (or Delirium), which is clinically distinct from the psychosis found in mental disorders such as schizophrenia or manic-depressive illness.

Although the terms insanity, dementia praecox, psychosis and schizophrenia have all been used to denote madness, psychosis and schizophrenia are clearly not synonymous. If being psychotic means to be hallucinated or delusional, this condition is always clinically transient and reflects a functional or "dimensional" state of the mind. On the contrary, clinical psychiatry views the loss of touch with reality as the inherent and fundamental characteristic of any psychotic condition. When a patient is delusional or hallucinated, he/she has lost touch with reality for two reasons: Because he/she has lost the capability of recognizing the boundaries between oneself and others and, more importantly, because he/she is not aware of such loss. Awareness of one's condition is permanently lost in schizophrenia, even after the patient has fully recovered from delusions and/or hallucinations. Clinical psychiatry only recognizes two syndromes in which the loss of touch with reality

S. Scarone (✉) • A. D'Agostino
Section of Psychiatry, Department of Medicine, Surgery and Dentistry, University of Milan Medical School, Milan, Italy
e-mail: silvio.scarone@unimi.it; armando.dagostino@unimi.it

N. Tranquillo (ed.), *Dream Consciousness*, Vienna Circle Institute Library 3,
DOI 10.1007/978-3-319-07296-8_31, © Springer International Publishing Switzerland 2014

combines with a continuous delusional or hallucinatory condition: Paranoia (or Delusional Disorder) and Chronic Hallucinatory Psychosis. Therefore, from a psychopathological point of view, the strongest similarity we can find between dreaming and psychotic mentation consists of the evidence that both lead the subject to a loss of awareness over their own condition (to be dreaming and, respectively, to be psychotic).

The available data are sufficiently robust and consistent to encourage us, as clinicians, to look forward to future clarifications on the neurofunctional basis of such absence of awareness in dreaming. It appears that the neurobiological profile of the dream state is characterized by a flattening of metabolism in frontal areas and a relative hyperactivity in limbic regions with a well-defined pattern of mutual relationships between the major neurotransmitters. Interestingly, many independent lines of research suggest similar neurofunctional patterns of modulation in psychosis.

The complexity of psychosis and its close relationship to distinctively human language-related cognitive domains limits the possibility of obtaining a satisfying animal model of this condition. The strength of a universally shared experience to which any observer may have direct access and the opportunity of correlating purely physiological findings to psychosis are some of the reasons why this approach seems worthwhile. At our current state of knowledge, it seems therefore reasonable to experimentally address the similarities and differences between dreaming and psychotic mental states. The protoconsciousness hypothesis appears to be a good starting point.

Chapter 32
You Emphasize the Continuity Between Waking and Dreaming. But What About Continuity in the Other Direction, i.e. Between Dreaming and Waking? And What About Discontinuity? Do You Deny Its Existence?

Michael Schredl

The three lectures of Allan Hobson cover a broad variety of topics regarding dreaming and its relationship to brain physiology and also its relation to waking life. The protoconsciousness hypothesis put forward by Allan Hobson postulates that one function of dreaming might be preparing consciousness for its tasks in waking life. Being a dream researcher for 20 years with a background in academic psychology and an advocate of the continuity hypothesis of dreaming, I would like to address a few issues about definitions and expand the review of the research findings regarding dream content because some of Hobson's claims are based on single studies and do not take into account the vast literature in the field.

I agree with Hobson when he writes that dreaming can be regarded as an altered state of consciousness and not as unconscious mental activity – as Sigmund Freud put it. The dream report – the only mean to the subjective world of dreaming – is a recollection of what was experienced during sleep. Although Hobson writes that most of the information in the brain is non-conscious in both states (waking and dreaming), e.g., control breathing as a simple but crucial activity of the brain, he uses the analogy that REM sleep and dreaming are two sides of the same coin (REM sleep dreaming to refer to a unified state of the brain-mind). This analogy is misleading because one side of the coin is much bigger than the other side; the brain is doing a lot more things during sleep than dreaming. Thus, equating the functions of REM sleep with possible functions of dreaming is not possible. Take for example the finding that the fetus in the last trimester and the newborn child spends about 50 % of the total sleep time in a sleep stage very similar to REM sleep in adults. If the function of this large amount of REM sleep is restricted to a cellular level, for example, the formation of synapses, dreaming – if some kind of subjective experience can be assumed – is not necessarily related to that function or even totally unrelated. A recent study (Schredl and Erlacher 2010), for example, showed

M. Schredl (✉)
Central Institute of Mental Health, Mannheim, Germany
e-mail: Michael.Schredl@zi-mannheim.de

N. Tranquillo (ed.), *Dream Consciousness*, Vienna Circle Institute Library 3,
DOI 10.1007/978-3-319-07296-8_32,

that task-related dream content is not correlated with performance gain in a procedural memory task trained in the evening and retested in the morning. The idea is the same; dreaming as subjective experience must not be related to possible long-term potentiation going on at a cellular level which is necessary to improve daytime performance. As the research addressing the relationship between dreaming and sleep-dependent memory consolidation is scarce, a conclusive statement cannot yet be made.

I personally do not like the psychiatric terms hallucination and delusion in reference to dreaming. In medical science, these terms are used for waking persons who hear or see things that are not there. Normally, a schizophrenic patient who hears voices is aware that they are not real, so this is not comparable to dreaming. The statement that a sleeping person hallucinates his or her dreams can only made by an outside observer but that's somewhat paradoxical because the dream is not visible to this observer and could only be obtained by waking the person up. But these are in my opinion problems of definition. The more interesting issue raised by Hobson is the question of how dreaming consciousness is different to waking consciousness. The fact that reading, writing, and working on a computer occurs in dreams less often than in waking life has been shown by several studies (Schredl and Hofmann 2003). This observation gives rise to the idea that the dreaming brain is not as capable as the waking brain in carrying out these cognitive skills. In fact, reading even a page of a book without any changes of the words and sentences has not been reported in the dream literature.

With respect to dream emotions, Hobson stated continuity between waking and dreaming in a sense that the dream emotion is very often consonant with the dream plot, e.g., while falling from a skyscraper you experience panic. On the other hand, Hobson states that the range of feelings are restricted to fear, anger, and elation whereas "social" emotions like sadness, embarrassment, and guilt are very rare. Even Hobson tells us that he is often alone in his dreams. This is not the case in young student samples. Social interaction with spouses, friends, and other people are an important ingredient of dreams (Hall and Van de Castle 1966; Schredl and Hofmann 2003). Another claim of Hobson that thinking is rare in dreams, especially autocriticism defined as an awareness of one's social impact upon another person, is not supported by research. Given the fact that persons tend to focus on the dream action and – if asked – on the dream emotions, the occurrence of thinking is dramatically underrated in spontaneous given dream reports (Meier 1993). In my own dreams, thinking is as prominent as in waking life, for example, what to do next, what other people might do, what might happen next. A good example is the typical dream of being inadequately dressed or naked in a public setting. This dream is full of thoughts about what other people think of one, embarrassment being the major feeling. As I said before, in this area, Hobson's review is limited to very few studies. Future research should look with greater care into the activity of the dreaming mind which is – in my opinion – capable of all second-order consciousness processes. This will shed light on the differences between dreaming consciousness and waking consciousness.

Hobson briefly reviews the continuity hypothesis of dreaming (Schredl 2003). Unfortunately, he did not refer to the broad literature showing that waking-life events affect dream content in regard to his protoconsciousness theory. The cited dream lag effect (Nielsen and Powell 1992) is not supported by the majority of the studies (cf. Schredl 2003). And I also wonder why he wants the continuity theorists to acknowledge that continuity is a two-way street because there is literature on the effect of dreams on subsequent waking life (e.g., Schredl and Erlacher 2007). Even more interesting is the second-order continuity reported by Schredl and Reinhard (2009–2010); dreams that were strongly affected by the previous day do affect in turn the mood of the next day most profoundly – even if emotional intensity of the dream is controlled for. These data suggest that the continuity between waking and dreaming is a two way street. Unfortunately, Hobson did not reflect on the significance of the effect of waking life on dreaming in his protoconsciousness hypothesis. The claim that continuity theorists do not "believe" in physiology is also difficult to understand because most modern psychological dream researchers – including myself – are interested in body-mind interaction, i.e., how close are dream features or actions related to brain physiology (Erlacher and Schredl 2008). As the first studies correlating images of the brain with dream content (Spoormaker et al. 2010) have been carried out, one can expect more conclusive findings in this area in the future.

Lastly, I would like to challenge Hobson's claim: "I have never learned anything from a patient's dream that I didn't already know." That working with dreams is effective has been shown by the Clara Hill group (Hill and Spangler 2007) and – doing some dream work myself – I would like to say that the aim underlying Hobson's statement is not correct. Instead, it should be: "Has working with the dream helped the patient to learn something about himself/herself which s/he did not know before?"

To summarize, Allan Hobson's three lectures are very stimulating to the field. I agree with his belief that there is a lot to be done in all domains of dream research, neurophysiology, psychology, and therapeutic areas.

References

Erlacher, D., & Schredl, M. (2008). Do REM (lucid) dreamed and executed actions share the same neural substrate? *International Journal of Dream Research, 1*, 7–14.

Hall, C. S., & Van de Castle, R. L. (1966). *The content analysis of dreams*. New York: Appleton-Century-Crofts.

Hill, C. E., & Spangler, P. (2007). Dreams and psychotherapy. In D. Barrett & P. McNamara (Eds.), *The new science of dreaming – volume 2: Content, recall, and personality correlates* (pp. 159–186). Westport: Praeger.

Meier, B. (1993). Speech and thinking in dream. In C. Cavallero & D. Foulkes (Eds.), *Dreaming as cognition* (pp. 58–76). London: Harvester Wheatsheaf.

Nielsen, T. A., & Powell, R. A. (1992). The day-residue and dream lag effects: A literature review and limited replication of two temporal effects in dream formation. *Dreaming, 2*, 67–77.

Schredl, M. (2003). Continuity between waking and dreaming: A proposal for a mathematical model. *Sleep and Hypnosis, 5*, 38–52.

Schredl, M., & Erlacher, D. (2007). Self-reported effects of dreams on waking-life creativity: An empirical study. *Journal of Psychology, 141*, 35–46.

Schredl, M., & Erlacher, D. (2010). Is sleep-dependent memory consolidation of a visuo-motor task related to dream content? *International Journal of Dream Research, 3*, 74–79.

Schredl, M., & Hofmann, F. (2003). Continuity between waking activities and dream activities. *Consciousness and Cognition, 12*, 298–308.

Schredl, M., & Reinhard, I. (2009–2010) The continuity between waking mood and dream emotions: Direct and second-order effects. *Imagination, Cognition and Personality, 29*, 271–282.

Spoormaker, V. I., Czisch, M., & Dresler, M. (2010). Lucid and non-lucid dreaming: Thinking in networks. *International Journal of Dream Research, 3*, 49–51.

Chapter 33
How Does the Finding of a Correlation Between the Three Conscious States (REM Dream, Lucid Dream, and Waking) and 40 Hz Power Fit with Your Suggestion That 40 Hz Is a Substrate of Consciousness?

Wolf Singer

33.1 The Putative Role of Synchronized Gamma-Oscillations in Brain States Associated with Conscious Experience

There are two distinct states during which brains are capable of having conscious experiences: attentive wakefulness and dreaming. The latter state appears to comprise two distinct forms, the normal dreams that are often recalled after awakening and the lucid dreams that are experienced as dreams while dreaming. These can apparently be controlled by the dreamer, thus representing an intermediate state between dreaming and wakefulness.

For cognitive contents to be consciously processed during wakefulness it is required that they are selected by attentional mechanisms. Recent evidence provides strong support for the hypothesis that attentional selection is achieved through the synchronization of neuronal responses in the gamma frequency range (Fries et al. 2001; Lima et al. 2011). This synchronization has the effect to enhance the saliency and propagation of the selected responses because the target neurons are driven more effectively by synchronous rather than temporally dispersed excitatory input. Moreover, communication between sending and receiving cell populations is selectively facilitated if both engage in gamma oscillations of similar frequency and phase lock at appropriate phase angles. Such is the case when the sent volleys arrive in the receiving, oscillating cell population at precisely those moments at which these cells are at the peak of their excitability cycle. These gating functions of phase adjusted oscillatory activity are particularly effective in the gamma frequency range

W. Singer (✉)
Max Planck Institute for Brain Research, Frankfurt Institute for Advanced Studies (FIAS), Ernst Strüngmann Institute (ESI) for Neuroscience in Cooperation with Max Planck Society, Frankfurt am Main, Germany
e-mail: singer@mpih-frankfurt.mpg.de

N. Tranquillo (ed.), *Dream Consciousness*, Vienna Circle Institute Library 3,
DOI 10.1007/978-3-319-07296-8_33,

because these high frequency oscillations allow for very precise adjustments of spike times and of the windows of opportunity (Fries 2005).

EEG and MEG analyses of the neuronal underpinnings of conscious processing – as compared to subconscious or subliminal processing – suggest that access to conscious processing requires transient synchronization of distributed oscillatory activity in the beta and gamma frequency range. Subconscious processing is associated with prominent gamma oscillations in the respective sensory and executive cortical areas but these activities seem to have access to consciousness only if they engage in a globally ordered state that is characterized by transient and precise phase locking (Melloni and Singer 2010; Melloni et al. 2007). Once this threshold is reached, the processed contents have access to working memory and eventually can undergo long term storage. Maintenance in working memory is associated with long range synchronization of theta oscillations which in turn are phase locked with bursts of gamma oscillations that occur during the peaks of the theta oscillations.

Unfortunately, it is very difficult to determine exactly when subjects are dreaming while asleep. Evidence from dream reports after awakening in different sleep stages suggests that dreams occur both in slow wave and paradoxical sleep. In the latter state, gamma oscillations are readily distinguishable. The electrographic signatures of this sleep stage are barely distinguishable from those of wakefulness. By contrast, gamma oscillations tend to be little expressed in states dominated by low frequency oscillations such as drowsiness or deep sleep. Analysis of cortical states with recording techniques allowing for high temporal and spatial resolution has revealed that the slow delta oscillations prevailing during deep sleep reflect periodic alternations between up- and down-states of large cell assemblies. During the up-states, the cells are strongly depolarized and engage in synchronized bursts of gamma activity. Thus, synchronized gamma oscillations could also serve as a read out mechanism of dream contents during slow wave sleep. As in this case, gamma activity is simultaneous with the delta rhythm. One would perhaps expect a difference in dream structure in comparison to dreams occurring during paradoxical sleep.

As far as lucid dreams are concerned, it seems as if they occur during a state of activated EEG and hence in a phase where gamma oscillations are readily observable.

In conclusion, all dream states are compatible with the occurrence of gamma oscillations. This supports the hypothesis that synchronized gamma oscillations have a critical role in processes that go along with conscious recollection of cognitive contents.

References

Fries, P. (2005). A mechanism for cognitive dynamics: Neuronal communication through neuronal coherence. *Trends in Cognitive Sciences, 9*(10), 474–480.

Fries, P., Reynolds, J. H., Rorie, A. E., & Desimone, R. (2001). Modulation of oscillatory neuronal synchronization by selective visual attention. *Science, 291*, 560–1563.

Lima, B., Singer, W., & Neuenschwander, S. (2011). Gamma responses correlate with temporal expectation in monkey primary visual cortex. *The Journal of Neuroscience, 31*(44), 15919–15931.

Melloni, L., & Singer, W. (2010). Distinct characteristics of conscious experience are met by large-scale neuronal synchronization. In E. Perry, D. Collerton, F. LeBeau, & H. Ashton (Eds.), *New horizons in the neuroscience of consciousness* (pp. 17–28). Amsterdam: John Benjamins.

Melloni, L., Molina, C., Pena, M., Torres, D., Singer, W., & Rodriguez, E. (2007). Synchronization of neural activity across cortical areas correlates with conscious perception. *The Journal of Neuroscience, 27*(11), 2858–2865.

Chapter 34
Do You Agree That Freud's Dream Theory Was Erroneous with Respect to (1) The Genesis of Dreaming (Release of Repressed Infantile Wishes) and (2) The Bizarreness of Dreams (Disguise and Censorship of Unacceptable Wishes)?

Mark Solms

It is clear from the present essay – notwithstanding the disclaimer at the outset – that Allan Hobson is still deeply preoccupied with disproving Freudian dream theory. What makes this preoccupation especially interesting is the fact that Hobson's models of dreaming have recently come to resemble Freud's. I would like to use this commentary to demonstrate how similar the two models now actually are. My aim in doing so is to clarify the last remaining point of difference. For the avoidance of doubt, I shall quote extensively from Hobson's essay.

Both models start from the observation that the psychological features of dreaming "are so distinctively different from those of waking as to suggest a major reorganisation of underlying brain activity." The major reorganisation is brought about by the state of sleep. This state entails, firstly, according to both Freud and Hobson, a withdrawal of engagement with the external world, but more importantly – and here too Hobson's model now agrees with Freud – it entails deactivation of "the dorsolateral prefrontal cortex (DLPFC), the seat of so-called executive ego mechanisms." This deactivation of the executive ego has very significant consequences for the sleeping mind, for the reason that "the dorsolateral prefrontal cortex plays a key role in keeping one on track on whatever state of consciousness prevails." Foremost amongst these consequences is a major reduction in "working memory, self-reflective awareness, volition and planning" – that is, a major reduction in what Freud called 'secondary process' cognition. This in turn releases the 'primary process' from normal inhibitory constraints.

> While dream consciousness may appear more free (or less constrained) than waking consciousness, it is not at all clear that dreamers 'decide' to do anything or exercise even

M. Solms (✉)
University of Cape Town, Cape Town, South Africa

Arnold Pfeffer Center for Neuro-Psychoanalysis, New York Psychoanalytic Institute, New York, NY, USA
e-mail: Mark.solms@uct.ac.za

N. Tranquillo (ed.), *Dream Consciousness*, Vienna Circle Institute Library 3,
DOI 10.1007/978-3-319-07296-8_34,

> illusional free will in any way . . .Things just happened to them, spontaneously as it were. They did not have the dream as much as the dream had them! (Hobson, Lecture II)

This is not the only respect in which Hobson's 'protoconsciousness' resembles Freud's 'primary process.' Protoconsciousness is, of course, also hallucinatory:

> According to my protoconsciouness hypothesis, in dreaming we positively activate the visual system and see, at the same time that we simultaneously deactivate analytic systems and do not think . . .The scientific study of dream consciousness shows very clearly that the brain-mind can be internally activated even when it is actively cut off from sensory input or motor output. When this occurs, internally generated signals drive the system in such a way as to make us believe that we are seeing. This insight strongly constrains our ideas not only about dream consciousness but also those ideas that we harbor about waking consciousness. From the subjective evidence of dreaming we can conclude that formed perceptions, the erroneous assumption that we are awake, and a narrative or scenario structure rich in social, motoric, and emotional content are all synthesized by the brain-mind itself. That capabilities for such an experience are given by brain physiology is also made clear indicating that the brain-mind, working on its own, creates a model of the world of truly amazing similitude to the world itself. This is primary consciousness. (Hobson, Lecture II)

Let us now review the functional properties of this "built-in consciousness generation system" which is released from executive ego control during sleep. First and foremost, we are told by Hobson that it is a "predictor" which "fills in" the brain-mind's "expectations." The brain-mind's intrinsic tendency to activate "its own expectant codes that are generated offline, via the spontaneous activation of neurons by internal signals" is normally constrained by realistic cognition; but in sleep "these critical cognitive functions are in abeyance." In consequence, the brain's primary "predictive model [is not] checked against reality" (ibid). This is precisely what Freud claimed: dreaming is characterised above all by the release of reality constraints upon *wishful* thinking.

I am well aware that Hobson will insist that his view of the brain's spontaneous tendency to generate "predictions" and "expectations" differs from Freud's view that it generates "wishes," so let us quickly move on to consider Hobson's view of the nature and origin of this primal mental tendency. He explains that the "ability to elaborate upon visual stimuli, to fill in, to flesh out, and to imagine the impossible, to create," which characterizes protoconsciousness and dreaming is due to the fact that:

> when dreaming, you are . . . psychotic . . . you have hallucinations and delusions. It is faint comfort to regard these processes as non-conscious [in waking cognition], as if that designation helped you escape the obvious: the hardware for psychosis is built into the brain. (Hobson, Lecture II)

It is generally recognized that the "hardware for psychosis" revolves centrally around the mesocorticolimbic dopamine system – the brain's intrinsic "predictive" and "expectant" system. It is therefore very important to note that Hobson now concedes that dreaming sleep is characterized by enhanced, pulsatile dopamine release. He still tends to minimize dopamine somewhat by mainly emphasizing its motor over its motivational functions, and likewise emphasizing the concomitant acetylcholine enhancement and demodulation of the other monoamine systems, but this does not detract from the fact that he now formally acknowledges the central role of dopamine in dream consciousness:

> Acetylcholine (ACh) and dopamine [DA] release are not suppressed in REM sleep. If anything, they may be more potent than they are in waking ... Dopamine and acetylcholine are both major players in normal motor control. It is thus plausible that the offline running of motor programs, emphasized by the activation-synthesis theory of dreaming, is supported by the abundant availability of these two neuromodulators in REM sleep. The pulsatile release of ACh and DA in the thalamus and cortex, and other subcortical motor areas, could be related to the subjective experience of motoric animation in REM sleep dreams. The fact that experimental animals with pontine lesions, and patients with REM sleep behavior disorder (RBD) act out their dreams is significant ... The net effect, in both cases, is a tendency for motor commands, normally suppressed in REM sleep by active motor inhibition, to be acted out, often awakening the animal or human subject. (Hobson, Lecture I)

Hobson does elsewhere in his essay also explicitly acknowledge that the "motor commands" issued during REM sleep are, above all, *instinctual* motor commands:

> [Animals in REM sleep with the above-described lesions] jumped to their feet and executed a complex motor sequence of attack and defense postures. An inescapable inference was that during REM sleep dreams such movements are commanded but they are normally prevented from expression by the motor inhibition ... The implications for the motoric animation of dreaming are self-evident. Protoconsciousness theory welcomes the functional significance of what would appear to be elaborate behavioral programming of instinctual acts such as escape, defense, and attack. The readout of these behaviors during REM sleep-without-atonia would seem to suggest that the brain is automatically programmed to emit important survival movements. These movements may normally be inhibited but can be experienced by us in dream consciousness ...This aspect of the theory fits with other predictive or anticipatory features of our virtual and actual behavioral repertoires. (Hobson, Lecture I)

In short, Hobson now admits that *dreaming consists in a primitive type of thinking, characterized by spontaneous "expectations" (of hallucinatory intensity), coloured by instinctual action tendencies that are normally suppressed and controlled by the "executive ego."* This might naturally lead to the assumption that dreams reveal primitive motivations. It is therefore not entirely surprising to find that, even in this respect, Hobson appears to be tentatively coming round to the Freudian view:

> Dream cognition, however bizarre, is usually salient with dream emotion. This salience is robust enough to lead us to consider that dream emotions may shape, or even trigger, dream cognition. Dreaming may be 'crazy' in its hallucinations and delusions, its incongruity and its discontinuity, but it does make sense emotionally. This could be good news for the dream interpreters, but a word of caution is indicated. The emotion dream-cognition link may be neither as unique nor as profound as most psychological dream theories suppose. Furthermore dreams may not be *uniquely* informative about an individual's associations to emotional stimuli. (Hobson, Lecture I)

So, although dreams are not *unique* in this respect, they do in principle admit of interpretation. This is because "the emotional-cognitive system usually, but not always, plays by a strict set of associative rules" (Hobson, Lecture I). Hobson elaborates:

> The unfettered play of dopamine in REM sleep is in keeping with the assumption that dreaming is "motivated" and that important motivational goals may be revealed in dreams. Dreams are not entirely meaningless and there is still room for considering them as uniquely hypermeaningful. This [however] is an unproved hypothesis... (Hobson, Lecture I)

Ever cautious, Hobson continues:

> Dynamically repressed (or actively forced down) mental content may well emerge in the process of dream image creation and plot selection processes that activation-synthesis credits with dream production, but such material is neither necessary nor sufficient for dreaming to occur in sleep. Dreaming is neither the only, nor necessarily the most privileged, way of getting at those psychic residues of trauma and conflict that constitute a kind of informational infection of the brain-mind. (Hobson, Lecture II)

Notwithstanding the ongoing disclaimers, he concludes:

> Sometimes dreams do reveal that earlier life issues, long believed dead, are still very hot in our non-conscious brain-minds. Freud does deserve credit for insisting on the long term persistence of conflict and trauma. Unjustified by dream science, however, is any interpretation scheme based upon the central Freudian hypothesis that forbidden infantile wishes cause dreams via the stimulation of disguise-censorship mechanisms. (Hobson, Lecture II)

This brings us to the only really substantive point of disagreement that remains between Hobson and Freud:

> The purpose of this shift [in sleep] from secondary to primary consciousness is not to protect secondary consciousness. On the contrary, it is to restore the sensitivity and specificity of dream consciousness to support waking consciousness. This specific example of brain activation shows how close the psychoanalytic and activation-synthesis models can get to one another; but the mechanisms and functions detailed by the two models are diametrically opposite. We are forced to choose between one or the other. Psychoanalysis views dreaming as unconscious mental activity that is designed to protect consciousness from disruption by the unconscious mind. Activation-synthesis regards dreaming as evidence of a built-in consciousness generation system. (Hobson, Lecture II)

In other words, for Hobson, *there is no conflict* between the sleeping executive ego and the instinctual action tendencies that are released during sleep. To be clear: while Hobson has moved "closer to Freud's idea of the ego ... in my use of this concept I am hoping to explain what the post-Freudian psychoanalysts called the 'conflict-free' ego." For this reason, according to Hobson, the anxiety that is so ubiquitous in dreams is simply a read-out of limbic anxiety-generating mechanisms; it is not a defensive response by the sleeping ego to the disinhibited limbic system. There is therefore no need to postulate the censorship function that (according to Freudian theory) reduces anxiety and thereby prevents the dreamer from waking up.

> It is inconceivable that the brain-mind states of wake and dream consciousness are not somehow complimentary and reciprocal ... It remains to be determined exactly how. We already know enough to make intelligent hypotheses of what this functional reciprocity might be. They include the enhancement and revision of learning and memory, the refreshment and maintenance of temperature control networks and the balance of circuits necessary to assure psychic equilibrium. These surprising functional benefits outweigh and diminish the psychoanalytic idea that dreaming is the guardian of sleep ... Freud's sleep

guardian theory is replaced by the physiological processes that guarantee the preservation of sleep in the face of internal brain activation. (Hobson, Lecture II)

In conclusion, then, although Hobson now accepts "as if we didn't know it already" that the available evidence proves "beyond the shadow of a doubt that dreaming is as much a psychodynamic as it is an organic [process]," he still finds no evidence for Freud's hypothesis to the effect that the function (or biological purpose) of the psychodynamics of dreaming is the preservation of sleep.

Fortunately, this last remaining point of disagreement – stripped of all the ideology – can be resolved empirically. I would like to propose a critical experiment to decide the issue once and for all. If Freud's sleep-protection hypothesis is correct, then patients who lose the capacity to dream (in consequence of posterior cortical lesions) should suffer poor quality sleep, with frequent micro-arousals, arousals and awakenings. If this is not found to be the case, then Freud's sleep protection theory will have been refuted. I accept responsibility for undertaking this critical experiment, and declare that I shall accept a negative result as definitive falsification of Freud's hypothesis. I hope that Hobson – as a truth-loving scientist – will likewise accept the corollary implications of a positive result.

Chapter 35
The Term "Hypnosis" Suggests a Sleep-Like State and Recent Evidence Regarding Lucid Dreaming Bears on the Notion of a "Hidden Observer." What Is Your View of the Analogy of Dreaming and Hypnosis?

David Spiegel

That Hobson's grand reformulation of the meaning of dreaming to consciousness has been greeted with hostility as well as approbation is testimony to its vigor. "To sleep, perchance to dream – aye there's the rub," quoth Hamlet, and rubbing generates friction. The idea that dreaming is a kind of warming up of consciousness for the demands of the day is consistent with the fact that REM sleep usually does not start until 90 min after falling asleep, and REM periods increase in length from 10 min at the beginning of the sleep cycle to an hour toward morning. Loss of REM sleep adversely affects daytime consciousness more than does loss of nonREM sleep. Depriving animals of REM sleep for several weeks leads to their death. Salient to Hobson's theory is the fact that the percentage of sleep devoted to REM activity is highest in infancy and childhood, even controlling for the fact that infants and children sleep longer than do adults. So as consciousness is growing and developing, REM activity is stirring it up. So perhaps dreaming is something akin to running the engine to start a balky car when the engine is cold – revving it up to catch and run normally.

Growing evidence of neural plasticity in the brain well into adult life also provides support for Hobson's choice. Neurons that fire together wire together, and our rich experiences during a given day trigger masses of dendritic growth and connection. Thus the problem is how do we keep our brains from outgrowing their skulls, and impeding processing with outmoded neural connections? One theory about sleep is that it is a time for neural pruning while sensory input is reduced. So the brain may be disposing of the day's residue of unnecessary synapses while preparing for the next day's processing. As Hobson notes, the hippocampus is storing memory in the cortex during slow-wave sleep, offloading information from the part of the brain that stores and retrieves it. That sensory deprivation is a *sine qua non* for sleep and dreaming is evidenced by the fact that a fairly low threshold of sensory input, like an alarm clock, will put an end to sleep.

D. Spiegel (✉)
Stanford University School of Medicine, Stanford, CA, USA
e-mail: dspiegel@leland.stanford.edu

N. Tranquillo (ed.), *Dream Consciousness*, Vienna Circle Institute Library 3,
DOI 10.1007/978-3-319-07296-8_35,

Hobson raises interesting analogies between dreaming and hypnosis: "It is walled off from conscious access, like certain aspects of mental activity that are made unavailable to waking consciousness by a hypnotist." The single-mindedness of dreaming is indeed like hypnosis. Hypnosis used to be called 'monoideaism.' It is a state of highly focused attention, with a reduction in peripheral awareness. While I have for decades emphasized the distinction between hypnosis and sleep, despite the common root word 'hypnos' in the name, Hobson's lectures provide an interesting connection between sleep and hypnosis. While the hypnotized person is clearly alert and awake, the hypnotic state of consciousness is, like in his description of dreaming, less self-reflective and critical. Indeed, there is also a suspension of critical judgment in hypnosis, akin to Hobson's lack of thinking in his bicycle-tree trimmer dream. In hypnosis, one does not think about what one is doing, hence the effectiveness of hypnosis in treatment (getting people to alter the perception of pain or the impulse to smoke). In hypnosis, you perform the behavior or change in perception rather than think about why you have not in the past, could not in the present, or will not in the future. Thus the power of hypnosis, even the existence of the phenomenon, may be owed in part to its similarity to dream state protoconsciousness.

Hobson's AIM model is fascinating, and he mentions its salience to hypnosis as well as sleep and dreaming: "If attention abates and comes to be directed by another person, then dissociated states, like hypnosis may be favored." I would differ a bit with his localization of hypnosis in the model. Attention is highly focused in hypnosis. I am making the assumption that dopamine is a key neurotransmitter driver of his activation construct. Dopamine is a key mediator of attention, frontally-based cognitive processing, reward, and, of hypnosis. We have found high correlations between levels of homovanillic acid, a dopamine metabolite, in the cerebrospinal fluid, and hypnotizability. Furthermore, several studies have found that a specific polymorphism, valine/methionine, in the catechol-o-methyl transferase (COMT) gene, is associated with higher hypnotizability than is found among those who are homozygotes (val/val or met/met). COMT is involved in dopamine metabolism, and the polymorphisms are associated with differences in figure/ground perception. Given this, I would place hypnosis in the near upper right portion of the cube – high activation (dopamine activity), high modulation (aroused, alert – high aminergic activity), but, and this is crucial, focused on internal rather than external inputs, with the exception of the hypnotist directing perception. I was once at the beautiful new baseball park in San Francisco, enjoying both the view of the game and of the Bay beyond. When I happened to gaze at the jumbotron broadcasting a televised view of the game, I realized that it took far less mental effort to let the director decide which portions of the game (and the environment) I should observe. No more choices of Bay or baseball diamond, pitcher or batter, first base or left field. One input. This is what the hypnotist provides – limited and structured access to input. So perceptual input from the outside world is reduced but not eliminated. The hypnotist becomes the director, restricting external perception.

Also, one of the most striking things about hypnosis is the subject's ability to alter perception – of pain, color vision, sounds. High hypnotizables can so alter lexical processing that they eliminate the Stroop color-word effect, which has been thought to be so automatic as to be uncontrollable. They can reduce or eliminate even severe pain. They can reduce blood flow in the lingual and fusiform gyri of the visual cortex in the direction of a hypnotically-suggested alteration in color vision. Hypnotically add color to a black and white grid, and blood flow increases. Drain the color and it decreases. Believing is seeing. Internally generated images compete with external perception. Hypnotized individuals respond to words but manipulate images. So hypnotized people play with perception rather than merely respond to it, which is in some ways, as Hobson points out, dreamlike.

Another neurophysiological analogy between trancelike states and REM sleep involves differential activation of anterior and posterior portions of the cingulate gyrus, positioned as it is between the limbic system (emotion) and the cortex (perception and cognition). It is interesting that the posterior cingulate is activated in REM, while the anterior cingulate gyrus is deactivated. The posterior cingulate is a prominent component of the default mode network, running when a person is awake but at rest and instructed not to perform a cognitive or perceptual task. The anterior cingulate is crucial to the salience network, active when a person is worried about solving a problem. So indeed, the posterior cingulate may be involved in 'daydreaming' as well as night dreaming.

Hobson's description of the brain during dreaming as being offline, actively processing internal information but shutting off perception of the outside world, is intriguing. This may in part account for his observation of the relative lack of self-consciousness: the internal is all there is, so there is no need to define self in contrast to the perceived world around, or perhaps no ability to do so. During dreaming, in Hobson's view, the self *is* the world, so no further definition is necessary.

Implicit in Hobson's description but not overtly stated is that dream consciousness proves how fallible our assessments can be. In dreaming, we think we see and indeed, as Hobson notes, have a beautifully constructed view of the world. But in fact we are fooling ourselves – thinking we see when we do not. It is rather a lesson in how credulous we are, and also in how much of perception is really an internally generated assembly of percepts and concepts.

I am a bit puzzled by Hobson's view of protoconsciousness as characterizing humans when asleep and lower mammals when they are awake. Clearly their ability to self-reflect and engage in propositional thought must differ, but there is a leavening effect of external perception and its impact on our perception that must affect lower mammals when they are awake in ways similar to our wakeful consciousness. We make the world understandable through internal prototypes and images, but the world makes us understandable by tempering our internally generated images, and more important, by forcing us to distinguish between self and other. While the aminergic uptick is a part of this shift to consciousness, so is the flood of external information. We may lack social emotion in dreams because we lack social input in sleep. It is not just the neural systems at play in a given mental state, but the content that they transmit, that shapes our level of consciousness.

I could understand some of the hostility of certain humanists and psychoanalysts to Hobson's description of dreaming were he to confine himself to neurobiological one-upmanship. But he does not – he artfully employs subjective description – both his own and that of many subjects whose dreaming he studies. He knowledgeably uses the neurobiology of sleep, but like a light rather than a club. His perspective should be sufficient to wake us up about consciousness and dreaming.

Chapter 36
How Does Your PHI Formula Deal with the Evidence that Consciousness Is State Dependent? More Specifically, if PHI Were Higher in REM Sleep Than in Waking, Would You Conclude That Dreaming Was More Conscious Even Than Waking?

Giulio Tononi

I like to say that dreaming is the single most important experiment in psychology and neuroscience. It proves that the adult brain, virtually cut off from the environment both on the input and the output side, can generate the entire world all of its own. That is, it gives rise to consciousness, and does it so well that at times dreaming experiences are hard to distinguish from waking ones. This fact is so fundamental that it forces us to acknowledge that consciousness is an *intrinsic* property of the brain, or at least of certain brain circuits, and to search for what this intrinsic property might consist of.

Dr. Hobson's question refers to the proposal that the relevant intrinsic property for consciousness is information integration (Tononi 2004, 2008). Information integration, expressed by the quantity (PHI), measures the extent to which a system is informationally more than the sum of its parts, implying that it cannot be reduced to them. High information integration requires the ability of multiple, specialized regions of the thalamocortical system to interact as a single entity (integration) while discriminating among a large repertoire of available states (information). In practice, a signature of high information integration in the thalamocortical system is that it should respond to perturbations with rapidly changing activity patterns (information) that affect a large portion of the cerebral cortex (integration) (Massimini et al. 2009). By contrast, information integration is low if the brain responds to perturbations in a stereotypical manner, or if the response remains local.

Recently, we tested some simple predictions of this hypothesis by comparing cortical responses to transcranial magnetic stimulation in combination with high-density electroencephalography (TMS/EEG) during wakefulness, sleep, and anesthesia. We found that in healthy awake subjects, TMS induces a sustained EEG response involving the sequential activation of different brain areas and affecting much of the cortex (Massimini et al. 2005), consistent with a high capacity for

G. Tononi (✉)
Department of Psychiatry, University of Wisconsin, Madison, Madison, WI, USA
e-mail: gtononi@wisc.edu

N. Tranquillo (ed.), *Dream Consciousness*, Vienna Circle Institute Library 3,
DOI 10.1007/978-3-319-07296-8_36,

information integration. By contrast, after loss of consciousness induced by general anesthesia, TMS pulses produced a stereotypic response that remained localized to the site of stimulation, indicating a breakdown of effective connectivity and thus of cortical integration (Ferrarelli et al. 2010). A similar breakdown of effective connectivity was observed during slow wave sleep early in the night (Massimini et al. 2005), when subjects report little or no conscious content upon awakening. Importantly, however, during rapid-eye-movement (REM) sleep, when subjects are unresponsive to sensory stimuli and virtually paralyzed but report vivid dreams upon awakening, the cortical response to TMS recovered its complexity and became similar to that observed during wakefulness (Massimini et al. 2010).

An ongoing effort is to try and quantify the results from such perturbation experiments to obtain a practical measure of the brain's capacity for integrated information that is related to the theoretical measure Φ. Such a measure would be especially useful to quantify consciousness during anesthesia as well as in brain-lesioned patients whose capacity for consciousness is hard to assess clinically (Alkire et al. 2008). In this context, Dr. Hobson asks what one should conclude if it turned out that Φ (or related measures) are higher in REM sleep than in waking. So far, all indications from TMS-EEG experiments are that, as one would expect from dream reports, information integration is high in wake, collapses in early NREM sleep, and recovers, though not fully, in REM sleep. One should keep in mind that, at present, the precision of our estimates of information integration is extremely crude, so it would be difficult to know what to make of small changes in estimated values of integrated information. Moreover, the level of consciousness is likely to fluctuate both within wake and within dreaming, so one would have to be careful in over generalizing. Nevertheless, it is clear that, if it turned out that information integration is considerably higher in deep NREM sleep or anesthesia than it is in wake or REM sleep, there would be something just as deeply wrong with the theory itself.

More importantly, it should be made clear that there is more to consciousness than its amount or quantity. Indeed, the integrated information theory states that each and every aspect of a given conscious experience is completely specified by an underlying, integrated information structure called a quale. That information structure is generated by the submechanism of a single informational entity or complex. In other words, it matters greatly not only *how much* information is integrated but also *how* precisely it is integrated. The analysis of dream reports indicated that dreaming consciousness, while remarkably consistent with many formal and content aspects of waking consciousness, often shows some peculiar differences. As pointed out by Dr. Hobson, such phenomenological differences are partly reflected in neurophysiological changes (Hobson et al. 2000). What we would expect is that, ultimately, the phenomenological features can be explained precisely in terms of the geometry of the corresponding information structures (Tononi 2008).

References

Alkire, M. T., Hudetz, A. G., & Tononi, G. (2008). Consciousness and anesthesia. *Science, 322*, 876–880.

Ferrarelli, F., Massimini, M., Sarasso, S., Casali, A., Riedner, B. A., Angelini, G., Tononi, G., & Pearce, R. A. (2010). Breakdown in cortical effective connectivity during midazolam-induced loss of consciousness. *Proceedings of the National Academy of Sciences of the United States of America, 107*, 2681–2686.

Hobson, J. A., Pace-Schott, E. F., & Stickgold, R. (2000). Dreaming and the brain: Toward a cognitive neuroscience of conscious states. *Behavioral and Brain Sciences, 23*, 793–842. discussion 904–1121.

Massimini, M., Ferrarelli, F., Huber, R., Esser, S. K., Singh, H., & Tononi, G. (2005). Breakdown of cortical effective connectivity during sleep. *Science, 309*, 2228–2232.

Massimini, M., Boly, M., Casali, A., Rosanova, M., & Tononi, G. (2009). A perturbational approach for evaluating the brain's capacity for consciousness. *Progress in Brain Research, 177*, 201–214.

Massimini, M., Ferrarelli, F., Murphy, M., Huber, R., Riedner, B., Casarotto, S., & Tononi, G. (2010). Cortical reactivity and effective connectivity during REM sleep in humans. *Cognitive Neuroscience, 1*, 176–183.

Tononi, G. (2004). An information integration theory of consciousness. *BMC Neuroscience, 5*, 42.

Tononi, G. (2008). Consciousness as integrated information: A provisional manifesto. *Biological Bulletin, 215*, 216–242.

Chapter 37
From Your Point of View as a Clinical and Health Psychologist, What Is Your Reaction to the Hypothesis of a Virtual Reality Program for the Brain?

Claus Vögele

The REM sleep–dream protoconsciousness hypothesis suggests that the brain states underlying waking and dreaming cooperate and that their functional interplay is crucial to the optimal functioning of both. Dreaming is proposed to be the subjective experience of a brain state with phenomenological similarities to – and differences from – waking consciousness, which is itself associated with a distinctive brain state.

Dream stimulation comes from the brain itself. These internal stimuli render dream consciousness more vividly perceptual than waking because of the exclusion of space-time data from the outside world. Cognitive processes (e.g. memory, self-reflective awareness, insight and judgment) are deficient in dreams because of a shift in balance from the aminergic to the cholinergic system. The dorsolateral prefrontal cortex (DLPFC) is deactivated in REM sleep.

These assumptions have important implications for the understanding of a range of mental disorders, including depression, some types of alcoholism, schizophrenia, eating disorders, borderline personality disorder, and other clinical conditions associated with short REM latency, increased REM sleep, increased REM sleep density, and loss of stages 3 and 4 sleep.

The neurophysiological concepts of the REM sleep–dream protoconsciousness hypothesis are consistent with the cholinergic-aminergic imbalance hypothesis of mood disorders, which proposes that depression is associated with an increased ratio of cholinergic to aminergic neurotransmission. The characteristic sleep abnormalities in depression may reflect a relative dominance of cholinergic activity, originating in the lateral dorsal tegmental and pedunculopontine tegmental regions, in relation to noradrenergic and serotonergic activity originating in the locus coeruleus, and the dorsal raphé nucleus, respectively. Antidepressant medications presumably reduce REM sleep either by their anticholinergic properties or by enhancing aminergic neurotransmission. Withdrawal from antidepressant drugs is

C. Vögele (✉)
Unité de recherche INSIDE, Université du Luxembourg, Walferdange, Luxembourg
e-mail: Claus.Voegele@uni.lu

N. Tranquillo (ed.), *Dream Consciousness*, Vienna Circle Institute Library 3,
DOI 10.1007/978-3-319-07296-8_37,

often accompanied by intense and prolonged dreams, which may reflect a REM rebound after drug-induced REM deprivation.

Dreaming allows for the organization of new experiences and their retention in long-term memory. This makes waking behavior more adaptive. These experiences can be emotional: over successive REM phases in one night, mood and emotion regulation is achieved by progressively containing surges of affective arousal.

Nevertheless, these processes can fail, most prominently in the case of anxiety dreams and nightmares, i.e. frightening dreams that are remembered only after waking up in the morning or from which the person wakes immediately, respectively. Frightening dream experiences are more common in children than adults; they also represent one of the major diagnostic features in a range of mental disorders including anxiety disorders (e.g. Post-traumatic Stress Disorder) and mood disorders. Both anxiety and mood disorders are characterized by arousal that causes REM sleep phases to be interrupted. This renders the process of dreaming incomplete and, therefore, the emotional processing unsuccessful.

If dreaming constitutes a virtual reality program for the brain then dreaming can be conceptualized as a process of fear extinction. Consequently, nightmares are an example of interrupted "exposure," which leads to the opposite of adaptation, i.e. sensitization.

Exposure is a crucial technique in cognitive-behavior therapy (CBT). It involves the repeated confrontation with a feared situation, object, thought or memory. From an emotional processing perspective, exposure leads to a belief change which presents new information that is incorporated into the patients' fear memory structures so that the strength of association between innocuous stimuli (e.g., feeling one's heart beating) and fear responses is reduced and the stimuli therefore no longer trigger anxiety. The effects of exposure therapy are also interpreted within the conceptual framework of extinction of learned fear responses. Specifically for panic disorder, enhanced resistance to extinction of fear conditioning has been shown experimentally. There are several learning theories that are relevant to exposure treatments such as conditioning, preparedness theory, or modern learning theories. One mechanism whereby cognitions may affect learned fear responses is by changing conditioned stimulus-unconditioned stimulus (CS-US) expectancies. Exposure therapy can be interpreted as an intervention that changes CS-US expectancies, i.e., the expectation that something bad will happen if the individual encounters the feared stimuli.

By re-experiencing a threatening situation we allow for adaptation processes to occur, including cognitive changes, fear extinction, re-evaluation and re-sorting of threatening memories. Dreaming and exposure share vital components in these processes, and therefore, its neural substrates. These structures include the amygdala and the dorsolateral pre-frontal cortex.

If these assumptions are true, we should investigate the activity of these neural substrates across the spectrum of consciousness. After decades of successful research highlighting the fundamental differences between dream consciousness and waking consciousness, it is perhaps now time to focus on the commonalities and to conceptualize consciousness as a continuum in relation to psychological and physiological processes.

Chapter 38
Please Summarize Your Findings on the Dream Representation of Disability in Your Studies of Handicapped People. What Bearing Do These Data Have on Protoconsciousness Theory?

Ursula Voss

We have studied the dreams of non-handicapped and individuals with a congenital handicap. The handicapped sample included deaf mutes, paraplegics, and adults with malformations of ligaments caused by Thalidomide. Each participant was asked to share ten dreams with us and to answer specific questions pertaining to the dream self: Did they dream of themselves as handicapped? Did they move and speak in their dreams? Did they perhaps experience more intense sensorial perceptions than non-handicapped people, as Freud's wish fulfilling hypothesis would predict?

Results very clearly show that the dream self is unharmed, even if the trauma had occurred before birth. Sometimes, the dreamer incorporates a word that is related to the handicap, like "deaf-mute" or "wheelchair" into the dream. It is peculiar, however, that these words are detached from their waking function. Very rarely, for example, will a wheelchair occur in the dream image of a paraplegic. However, the dreamer is not seated in it, or if he is, he can get up from it easily to run or swim. Likewise, a deaf-mute person will sometimes dream of visitors from his deaf-mute circle of friends, but they can all talk and sometimes even speak in a foreign language on the phone. When we asked three psychologists and one non-psychologist to identify the dream reports of deaf-mutes, paraplegics and controls, they reached an inter-rater agreement of merely 26 %. It was not at all possible to make a blind match of dream to dreamer. Interestingly, the psychoanalyst was the one with the most liberal answering style and the most false assignments in the group of non-handicapped individuals.

We found no evidence of wish fulfillment in dreams, as paraplegics did not move more often in dreams than controls and deaf-mutes did not report more spoken conversations than controls. Experiences in other sensory modalities were also not different between groups. For example, a deaf-mute individual had equally intense

U. Voss (✉)
Universität Bonn, Bonn, Germany

JW Goethe-Universität, Frankfurt, Germany
e-mail: voss@psych.uni-frankfurt.de

N. Tranquillo (ed.), *Dream Consciousness*, Vienna Circle Institute Library 3,
DOI 10.1007/978-3-319-07296-8_38,

and equally frequent visual or tactile perceptions as a non-handicapped person. Based on these findings, I think it is reasonable to assume that a deficit in one sensory modality in waking is not compensated for by an overrepresentation of sensorial experiences in the dream. This speaks against the wish fulfillment hypothesis.

38.1 How Do the Data Fit with Protoconsciousness Theory?

Protoconsciousness implies a functionality of dreams, namely that it serves waking consciousness. One such proposed function is to practice virtual movements. The fact that our handicapped subjects were able to move just as much and as intense as non-handicapped and not be surprised or enthusiastic about it, suggests that we are accessing perhaps innate motor and perceptual programs in our dreams that remain unaffected by waking reality, just as suggested by protoconsciouness theory.

Our findings suggest that dream consciousness is qualitatively distinct from waking consciousness, at least with respect to the concept of self. While its determinants still have to be investigated, it seems that dream consciousness is both richer and poorer than waking consciousness. It is also evident that the dream self is at least partially independent of the wake self.

Dream consciousness is richer than waking consciousness in that the dream self is capable of engaging in behaviors that would be impossible in waking. Paraplegics who have never learned how to walk can move their legs without even noticing that this experience is unique. All of us, handicapped or not, know how it feels to fly, just like we are, sometimes using our hands and feet for balance and direction and sometimes not even that. In the dream, perception is internally generated. The feedback system with the external world that keeps us safe from jumping off dangerous heights, for example, becomes obsolete. It seems that the limits of dream imagery – including those images related to the self – are the boundaries of our minds. In the dream, we explore possibilities, it seems, not realities. In this "protoconscious" way, dream virtuality may prepare us for waking reality by creating these possibilities that may or may not be executable at first try, and some may not be possible at all, like for instance, using and feeling your legs when you have never done so in your waking life.

Dream Consciousness is poorer than waking consciousness because we lack reason, memory, and reflective insight. The dream rotates around the present, and where it connects to elements of our past and perhaps projects into the future, the networking is perforated and faulty, as evidenced by reduced cortico-cortical coherences in the EEG. Although the dream is not a state of decapitation, it is a state of little self-agency. We have yet to identify the physiological structures correlating with dream content.

Protoconsciousness theory is currently the best model to explain our findings on the unharmed self image of people with a congenital handicap. It lays the groundwork and the theoretical basis for rigorous experimental hypothesis testing.

Chapter 39
You Have Made Blind Sight a Valid Phenomenon. What Is Your Position About Dream Vision? Is That Not Evidence for Internal Visual Image Generation by the Brain? Do You Suppose That Such Image Production Could Occur in Blind Subjects?

Lawrence Weiskrantz

The typical blindsight subject has one half-field of INTACT vision, with which he can negotiate the world, etc. There is no way (of which I am aware) of deciding whether a visual dream emanates from the intact or the blind hemifield. It would be important to study patients with total blindsight over the entire field due to bilateral visual cortical damage. Such patients are very rare.

In studying one blindsight patient, we did have evidence of internal visual generation, in one sense. This was a subject who had vivid after-images of the stimuli in the blind field of which he was unaware. When we presented him with a red stimuli to one eye and a green stimulus to the other eye, which in normal subjects generates a "cortical yellow" appearance, this subject reported a BLUE after-image, i.e., the complimentary colour of yellow. He did not see cortical yellow as such. The outcome could not be due to bleaching at the retinal level because that would have generated yellow. And so the after-image had to be generated brain activity induced by the initial stimuli. This is reported in my recent book, *Blindsight* (2009).

Reference

Weiskrantz, L. (2009). *Blindsight: A case spanning 35 years and new developments* (2nd ed.). Oxford: Oxford University Press.

L. Weiskrantz (✉)
Department of Experimental Psychology, University of Oxford, Oxford, UK
e-mail: larry.weiskrantz@psy.ox.ac.uk

N. Tranquillo (ed.), *Dream Consciousness*, Vienna Circle Institute Library 3, DOI 10.1007/978-3-319-07296-8_39, © Springer International Publishing Switzerland 2014

Chapter 40
How Can the Protoconsciousness Hypothesis Contribute to Philosophical Theories of Consciousness and the Self?

Jennifer M. Windt

40.1 Dreams, Protoconsciousness, and Minimal Phenomenal Selves

In his William James lectures on Dream Consciousness, Allan Hobson provides a rich and multi-faceted introduction to state-of-the-art findings from sleep and dream research, as well as an overview of his own wide-ranging contributions to this field. It is no exaggeration to say that his work has shaped contemporary theories of sleep and dreaming like no other. But he has done even more: by showing how his work on dreaming fits into broader theories of consciousness (Hobson et al. 2000) and psychotic wake states (Hobson 1999), he has repeatedly emphasized its implications for the theoretical understanding of consciousness itself. His long-standing willingness to connect the fields of sleep and dream research with contemporary work on consciousness whilst bridging disciplinary gaps is of tremendous value, especially given the fact that dream research continues to be a fairly marginalized field. The success of this approach is reflected in the broad range of contributions from leading researchers in various fields in this volume. It will be of great value for scholars and students for years to come.

Continuing this tradition, the protoconsciousness hypothesis addresses an important desideratum for current dream research: What, if any, are the functions of dreaming? One of its strengths is that Hobson's answer to this question suggests that REM-sleep dreaming "provides a structural and functional building block for waking consciousness" and that the comparative study of dreaming and wakefulness is "a way of understanding not only dreaming but the evolution of consciousness itself."

In my commentary, I will distinguish a stronger and a weaker reading of the protoconsciousness hypothesis. I will then argue that even on the weaker reading,

J.M. Windt (✉)
Philosophisches Seminar, Johannes Gutenberg-Universität Mainz, Mainz, Germany
e-mail: windt@uni-mainz.de

N. Tranquillo (ed.), *Dream Consciousness*, Vienna Circle Institute Library 3, 225
DOI 10.1007/978-3-319-07296-8_40, © Springer International Publishing Switzerland 2014

the protoconsciousness hypothesis is a tremendously fruitful theoretical tool for understanding the structure of phenomenal experience in dreaming and wakefulness. Dreaming is the most wide-spread and robust example of phenomenal consciousness involving minimal phenomenal selfhood. To put this point in Hobson's terms, dreams are protoconscious in the sense that they are tied to the experience of a protoself. Because dreams exemplify the simplest form of self-consciousness, they present an important contrast condition for contemporary philosophical theories of consciousness and the self that are largely focused on standard wake states. Consequently, the example of dreaming also fulfills an important metatheoretical function for philosophical theorizing about self-consciousness.

On the stronger and primary reading of the protoconsciousness hypothesis, REM-sleep dreaming fulfills an *evolutionary* function in the development and maintenance of consciousness. Birds and mammals, though they evolved separately, all have REM sleep, are homeothermic, and have considerable behavioral and cognitive skills (Hobson 2009). Aside from widespread evidence that REM sleep has a restorative effect on thermoregulation, the preponderance of REM sleep during early developmental stages in birds and mammals (including humans) suggests that it may also prepare for the integrative functions of consciousness (Hobson, Lecture III). Even before the emergence of full-fledged dreaming, REM sleep might involve a "virtual world model, complete with an emergent imaginary agent (the protoself) that moves (via fixed action patterns) through a fictive space (the internally engendered environment) and experiences strong emotion as it does so" (Hobson 2009, p. 808). Moreover, the persistence of REM sleep throughout adulthood suggests that "it subsequently plays an equally indispensable part in brain–mind maintenance" (Hobson, Lecture III).

Assessing this stronger reading of the protoconsciousness hypothesis as a thesis about the *evolutionary* function of dreaming would require a detailed discussion of a number of questions. For instance, granting that adults' dreams typically have the structure of a protoself moving through a virtual world, is it permissible to assume that protoconscious REM sleep in other mammals and in early developmental stages in humans takes a similar form? What exactly does it mean to say that REM sleep is protoconscious even before the emergence of full-fledged dreaming? And to what extent does the protoconsciousness hypothesis really ascribe an evolutionary function to the conscious experience of dreaming, and not just to REM sleep? Discussing these questions in any detail would be well beyond the scope of this commentary. Luckily, as will become clear below, the philosophical value of the protoconsciousness hypothesis for theories of consciousness and the self can be assessed even if no answer to these questions is in place. For both reasons, I want to concentrate on a weaker reading.

On this weaker reading, the protoconsciousness hypothesis is about the *structure* of phenomenal experience in dreams rather than about its function. It predicts that dreams exemplify a simple and, in a sense, more fundamental form of conscious experience than that found in standard wake states, and one that involves the experience of a self—a protoself—in a virtual environment.

Note that this point is independent of the question of whether REM sleep dreaming additionally fulfills an evolutionary function, or which function exactly this is. Whatever the answer to this question turns out to be, it is compatible with saying that phenomenal experience in dreams—or at least in the dreams of adults, as studied through dream reports—is not just a "scenario generator" (Hobson, Lecture I), but generates a very specific type of scenario, namely one that is centered on a phenomenal self. Interestingly, though dreams preserve the general structure of standard waking consciousness by simulating the experience of a self in a world, important dimensions of self-experience are typically missing in dreams (Metzinger 2003, 2009; Windt and Metzinger 2007). A majority of dreams only involve a phenomenal self in a conceptually weak sense as compared to standard wake states, and in some dreams, the dream self is even reduced to the minimal conditions for phenomenal selfhood to arise (Windt 2010).

Note also that this is not a trivial suggestion. For instance, Hobson's dream of cycling through London with a tree trimming tool (Hobson, Lecture I) is much more complex. It is a good illustration of a number of the stereotyped formal features that characterize a majority of dreams. As Hobson and colleagues write elsewhere, these include:

> vivid sensorimotor imagery that is experienced as waking reality despite such distinctive cognitive features as impossibility or improbability of time, place, person and actions; emotions, especially fear, elation, and anger predominate over sadness, shame, and guilt and sometimes reach sufficient strength to cause awakening; memory for even very vivid dreams is evanescent and tends to fade quickly upon awakening unless special steps are taken to retain it. (Hobson et al. 2000, p. 795)

Such a stereotyped definition focusing on the formal features shared by a majority of dreams is helpful if the target is a comprehensive theory of dreaming. If, however, the goal is to understand the fundamental structure of conscious experience itself, then something simpler is needed. In particular, the emphasis will now no longer be on all of the formal features that characterize a majority of dreams, but on the *phenomenal core* of dreaming, that is, on those formal features exhibiting the *highest degree of statistical invariance* across different types of dreaming, but also standard and altered wake states (Windt 2010, 2014).

I have previously argued that dreaming essentially involves *immersive spatiotemporal hallucination*: the sense of immersion in a spatiotemporal reference frame, where this reference frame is created *offline* rather than online, as in perception. Importantly, some dreams involve a sense of self-location independently of modality-specific forms of imagery (e.g. visual, auditory, or motor imagery), thoughts, or emotions. These dreams are retrospectively described as having involved a self, but phenomenal selfhood takes the minimal form of pure self-location; in its simplest form, even the phenomenology of embodiment is lost, and the dreamer may even identify with a non-extended point in an amodally experienced spatial expanse (cf. LaBerge and DeGracia 2000; Occhionero et al. 2005). On the phenomenological level of description, the experience is simply that of being *here*, at this point in space, and *now*, at this point in time (Windt 2010, 2014).

This description of minimal phenomenal selfhood lacks many of the formal features that characterize a majority of dreams and presents a more parsimonious interpretation of Hobson's claim that dreaming involves a protoself. Where Hobson claims that "dream consciousness guarantees the binding of sense of self, motility, sensation and emotion" (Hobson, Lectures Introduction), on my weaker view, the sense of motility, sensation, and emotion can be lost. Obviously, this is essentially an empirical question. At the same time, my view preserves what I take to be the central point of the protoconsciousness hypothesis, according to which dream consciousness guarantees the binding of a minimal sense of selfhood—in terms of self-location—to a virtual environment, a phenomenal space. For this reason, I think it is compatible with Hobson's views on protoconsciousness and the protoself.

Once the protoconsciousness hypothesis is regarded as a thesis about the phenomenal structure of dreaming, it becomes clear that empirical research programs about dreams fulfill an important *metatheoretical* function in that they help isolate the simplest structure of phenomenal selfhood. Why? On a purely conceptual level, they advance our understanding of what all first-order theories describing the phenomenon of self-consciousness across a multitude of disciplines and with different conscious states as their target have in common. Taken together with findings from out-of-body experiences and full-body-illusions (Lenggenhager et al. 2007; Blanke and Metzinger 2009), dreams suggest that this structure is not state-dependent, but that minimal phenomenal selfhood, or the phenomenology of pure spatiotemporal self-location, can occur in altered wake states as well. Because spatiotemporal self-location can combine with the phenomenology of embodiment, thoughts, and emotions, we now have not just a positive thesis about the phenomenology of dreaming, but a conceptual claim about self-consciousness in general. Put in Hobson's terms, minimal phenomenal selfhood provides not only "a structural and functional building block" for more complex forms of self-consciousness, including those seen in standard wake states and in a majority of dreams, but also a theoretical tool, a *conceptual* building block for theories of self-consciousness. For philosophy of mind and interdisciplinary consciousness research, this is an important insight.

At this point, one might object that by ascribing a purely metatheoretical function to recent advancements in our understanding of dreaming, I am distracting from the original point of the protoconsciousness hypothesis. It was, after all, proposed as a first-order theory about the function of *dreaming*, and was not intended to describe the second-order function of more abstract, metatheoretical work on dreaming and its implications for philosophical theories of self-consciousness. And though there is some truth to this, note that I am not thereby denying that even on a stronger reading, the protoconsciousness hypothesis may have important contributions to make to theories of self-consciousness. I am only making the more modest claim that part of its genuine *philosophical* value—and perhaps, the most important part—stems from the weaker reading suggested here. Seen in this context, my deliberately weak reading of the protoconsciousness hypothesis is one with which Hobson himself can readily agree. He claims, for instance, that "dreaming reveals many fundamental aspects of how consciousness is created by the brain" (Hobson, Lecture III) and that "dreaming [...] should be seen as a key building block in any general theory of how the brain-mind works" (Hobson, Lecture III). The claim that the example of dreaming fulfills an important metatheoretical function is, after all, inseparable from the protoconsciousness hypothesis.

Again, this point is by no means trivial. In a field that is otherwise characterized by strong and persistent controversies, it is striking that the core claims of the protoconsciousness hypothesis—insofar as it is read as a thesis about the phenomenology of dreaming—are widely shared by leading representatives of dream research and philosophy of mind. For instance, Tore Nielsen has recently proposed that "that the subjective nature of dreaming consists of *a convincing simulation of waking reality experience*" (Nielsen 2010, p. 595). Antti Revonsuo (2006) has used dreaming to develop the "world-simulation metaphor of consciousness," emphasizing that dreaming, like wakefulness, consistently involves the experience of a self in a world. And Metzinger (2003, 2009) has used the example of dreaming to illustrate his broader claim that consciousness itself is a "global simulational state" and that the phenomenal self is, in an important sense, an illusion. This deep point of agreement suggests that aside from its metatheoretical contribution to theories of consciousness and the self, the protoconsciousness hypothesis might also fulfill an *integrative* and *dialectical* function in the discussion on dreaming.

References

Blanke, O., & Metzinger, T. (2009). Full-body illusions and minimal phenomenal selfhood. *Trends in Cognitive Sciences, 13*(1), 7–13.

Hobson, J. A. (1999). *Dreaming as delirium. How the brain goes out of its mind*. Cambridge, MA: MIT Press.

Hobson, J. A. (2009). REM sleep and dreaming: Towards a theory of protoconsciousness. *Nature Reviews Neuroscience, 10*, 803–813.

Hobson, J. A., Pace-Schott, E. F., & Stickgold, R. (2000). Dreaming and the brain. Toward a cognitive neuroscience of conscious states. *Behavioral and Brain Sciences, 23*, 793–842.

LaBerge, S., & DeGracia, D. J. (2000). Varieties of lucid dreaming. In R. G. Kunzendorf & B. Wallace (Eds.), *Individual differences in conscious experience* (pp. 269–307). Amsterdam: John Benjamins.

Lenggenhager, B., Tadi, T., Metzinger, T., & Blanke, O. (2007). Video ergo sum: Manipulating bodily self-consciousness. *Science, 317*(5841), 1096–1099.

Metzinger, T. (2003). *Being no one: The self-model theory of subjectivity*. Cambridge, MA: MIT Press.

Metzinger, T. (2009). *The ego tunnel: The science of the mind and the myth of the self*. New York: Basic Books.

Nielsen, T. A. (2010). Dream analysis and classification: The reality simulation perspective. In M. Kryger, T. Roth, & W. C. Dement (Eds.), *Principles and practice of sleep medicine* (pp. 595–603). New York: Elsevier.

Occhionero, M., Cicogna, P., Natale, V., Esposito, M. J., & Bosinelli, M. (2005). Representation of self in SWS and REM dreams. *Sleep and Hypnosis, 7*(2), 77–83.

Revonsuo, A. (2006). *Inner presence: Consciousness as a biological phenomenon*. Cambridge, MA: MIT Press.

Windt, J. M. (2010). The immersive spatiotemporal hallucination model of dreaming. *Phenomenology and the Cognitive Sciences, 9*, 295–316.

Windt, J. M. (2014). *Dreaming: A conceptual framework for philosophy of mind and empirical research*. Cambridge, MA: MIT Press.

Windt, J. M., & Metzinger, T. (2007). The philosophy of dreaming and self-consciousness: What happens to the experiential subject during the dream state? In D. Barrett & P. McNamara (Eds.), *The new science of dreaming* (Cultural and theoretical perspectives, Vol. 3, pp. 193–248). Westport: Praeger Perspectives.

Chapter 41
How Does the Formal Approach to Mind Taken by Protoconsciousness Science Compare with the Formal Approach to Works of Art That You Espouse?

Hellmut Wohl

On December 8, 1937, Sigmund Freud wrote to André Breton that:

> The superficial aspect of dreams, what I call the manifest dream, holds no interest for me. I have been concerned with the latent content which can be derived from the manifest dream by psychoanalytic interpretation. A collection of dreams without associations and knowledge of the context in which it was dreamed does not tell me anything, and it is hard for me to imagine what it can mean to anyone else.

The associations and the context are presumably the material that consciousness has repressed and relegated to the unconscious, and which can be reconstructed by psychoanalytic interpretation of its disguised emergence in the manifest dream.

According to Allan Hobson's hypothesis of dreaming as arising not from the unconscious but as an altered state of consciousness, or protoconsciousness, (what Freud called the manifest dream and its scenario construction) is generated automatically by the brain-mind during both REM and NONREM sleep. Hobson's approach to dreams is formal in the sense that it describes a scientifically based schema for understanding the process of dreaming. By formal is meant the mental, visual, and emotional characteristics of conscious experience, whatever its specific content. The evidence of neurophysiology strongly suggests that dream form is ontologically primary to dream content, and that any approach to dream content which attempts to explain dream form by the analysis or interpretation of dream content is erroneous.

The salient point in comparing Hobson's formal approach to dreams with a formal approach to works of art is that in art, too, form is primary to content. The study of art since the 1930s has been dominated by its definition as a symbolic language. The advantage of this definition is that it considers art as a system of signs and symbols connected with the symbolic systems of linguistics, philosophy, psychology, and anthropology. However, as a result of its virtually exclusive

H. Wohl (✉)
Boston University, Boston, MA, USA
e-mail: hellmutwohl@gmail.com

N. Tranquillo (ed.), *Dream Consciousness*, Vienna Circle Institute Library 3,
DOI 10.1007/978-3-319-07296-8_41,

attention on the study of meaning, another definition of art as a system of formal relations has been and continues to be neglected, even though it is self-evident that no meaning can be conveyed without form. The structural forms of art - solids and voids in architecture and sculpture, tones, lines and areas in painting – can be apprehended independently of meaning; and they are governed by more or less regular evolutions in time without relation to meaning. In *The Life of Forms in Art*, Henri Focillon developed the thesis that the formal language of art, or of style, has a morphological and evolutionary existence of its own which persists through a range of mutations until its inherent possibilities are exhausted and one style is replaced by another (Focillon 1989); and George Kubler in *The Shape of Time* has studied more specifically the morphological problems of the duration of styles in series and sequence (Kubler 1962).

The formal approach to art gained momentum in the early twentieth century through the work of the Viennese art historians, Franz Wickhoff and Alois Riegl (Wickhoff 1912; Reigl 1927), which replaced the moralizing judgment of "degeneracy" of late Roman art with the hypothesis that one style, or system of the formal language of art, was replaced by a different system of equal value, thus dividing the history of art, as Kubler has put it, "along structural lines marked by the frontiers between types of formal organization." At about the same time, a different concept of stylistic sequence was advanced by the Swiss art historian, Heinrich Wölfflin (Wolfflin 1950). By pointing out five polar opposites in the realization of form in Italian Renaissance and Baroque, Wölfflin characterized fundamental differences in morphology in the two periods. He did not, however, define the individual qualities, or the specific identity, of works of art.

The identity of a work of art: when, by whom, and how it was made is the objective connoisseurship, which became a systematic method of evaluating and identifying the authorship of works of art in the later nineteenth century. The criterion for determining a work's identity was succinctly described by the great scholar of early Italian painting, Richard Offner (Offner 1998). The history of art, Offner wrote:

> should be evolved directly out of its concrete examples [and thus] free the object of the verbal system with which the literature of art has overlaid it. For the nature of language is such that the object must surrender its physical attributes before it can be converted into words.. ...

There is a consilience between Offner's formal criterion and Hobson's formal approach to the study of dreaming. Hobson's liberation of dreams from their verbal overlay, Freudian, moralistic, or otherwise interpretative, is analogous to Offner's point of freeing "the object of the verbal system with which the literature of art has overlaid it."

One among many examples is a painting of 1939 by René Magritte in the Art Institute of Chicago with the title *Time Transfixed.* A locomotive is suspended in mid-air in front of a fireplace. On the shelf above it is a clock set at 12:44. The images in the painting are as strange and unsettling as the disquiet we experience in dreams. Magritte represents the bizarreness of dreams in a neutral, scientifically impersonal way. The locomotive (phallic) and the open fireplace (vaginal) would be an easy mark for Freudian psychoanalysis. But that would miss the formal point of the work: the incongruity of objects and setting such as that which we experience in dreams. Magritte himself thought of the painting as a metamorphosis. The locomotive, he wrote, is:

> Charging out of the chimney opening of the fireplace in a dining room instead of the usual stove pipe. This metamorphosis is called time transfixed.

The representation of time in the painting is a visualization of time in dreams. Because works of art, like dreams, are closed formal systems sealed off from input from without and not subject to the conditions of how we perceive or make judgments about the world around us, the juxtaposition of the locomotive charging out of the fireplace and a clock that has stopped at 12:44 are as paradoxical, and at the same time as believable, as the disorientation in respect to time that is characteristic of dreams. In a statement of his intentions, Magritte wrote that:

> Too often by a twist of thought we tend to reduce what is strange to what is familiar. I instead intend to return the familiar to the strange.

Which is exactly what happens when we are dreaming.

References

Focillon, H. (1989). *The life of forms in art*. New York: Zone Books.
Kubler, G. (1962). *The shape of time*. New Haven: Yale University Press.
Offner, R. (1998). *A discerning eye: Essays on early Italian painting*. University Park: Pennsylvania State University Press.
Riegl, A. (1927). *Die Spätrömische Kunstindustrie*. Wien: Staatsdruckerei.
Wickoff, F. (1912). *Römische Kunst (Die Wiener Genesis)*. Berlin: Meyer & Jessen.
Wölfflin, H. (1950). *Principles of art history: The problem of the development of style in later art*. Translated by M. D. Hottinger, New York: Dover Publications

Part III
Response to Commentaries on the William James Lectures on Dream Consciousness

Chapter 42
Lecture I: Psychology

J. Allan Hobson

42.1 What Is Dreaming?

Very little attention was paid by the commentators to the question of the definition of dreaming. This suggests enough consensus to guarantee broad agreement despite still lingering doubt about whether dreaming is a state of consciousness. Susan Blackmore disapproves of my calling dreaming a state of anything but the self. John Kihlstrom says that dreaming may well be an experientially conscious state but insists that it is of unconscious origin. I acknowledge the merits of both of these ideas. John Antrobus and Michael Schredl still seem to be unpersuaded that the formal features of dreaming are significantly different from those of waking. In raising these reservations, however, no one questioned that my report of subjective experience, occurring in my sleep, of riding a bicycle through London traffic with a long handled tree trimmer across the handlebars, was an account of dreaming. Nor did any one suggest that such a mental experience could occur even in fantasy. I conclude that all commentators recognize a dream report when they see one and they all know that such mentation does not occur in waking. That is all the consensus that I want at this point.

42.2 When Does Dreaming Occur?

Everyone agrees that dreaming occurs in sleep and the commentators at least tacitly admit that REM sleep is the brain state most powerfully associated with dreaming. There is actually more controversy around this claim than the current collection of commentators expresses. Many psychologists (including David Foulkes who was

J.A. Hobson (✉)
Department of Psychiatry, Harvard Medical School, Boston, MA, USA
e-mail: allan_hobson@hms.harvard.edu

N. Tranquillo (ed.), *Dream Consciousness*, Vienna Circle Institute Library 3,
DOI 10.1007/978-3-319-07296-8_42,

invited but declined to comment) dismiss the relevance of physiology to dream science because reports of dreams can follow arousal from NREM sleep. I have conceded to such critics that the correlation of dreaming with REM is not perfect. The correlation is quantitative, not qualitative. The AIM model takes probability into account and quantifies it as shown in Lecture II.

Tore Nielsen gives us a summary of his covert REM hypothesis which attributes NREM dreaming to an admixture of dissociated REM processes. Nielsen's theory is not unlike my view of REM physiology as suppressed in but not entirely absent from waking. For me the correlation between REM and dreaming is so powerfully quantitative as to encourage a focus on REM in striving to develop a scientific psychophysiology.

I was pleased by Dan Dennett's explanation of his only apparent commitment to Norman Malcolm's famous but now untenable contention that dreaming does not really occur in sleep but only occurs in the confusional moments of awakening and is then back projected into sleep. Dennett acknowledges that this theory, which was also made by Freud, is erroneous, and that he was only using it as a springboard for his philosophical purposes. Dreaming – or dream like mental activity – does really occur on awakening and it is often back projected into sleep but it does not only occur at this time. In fact, dreaming is far more plentiful and more robustly bizarre within sleep where it is correlated with the totally unanticipated brain activation of REM discovered by Aserinsky and Kleitman in 1953. Numerous empirical studies, including those of my own lab, have documented the correlation of dreaming and REM (See Hobson et al. 2000, for a review). Philosophers may quarrel with my assertion that such dreaming is a unique and tellingly different state of consciousness but I am pleased to note that so many commentators agree with the central tenet of my new protoconsciousness theory.

Commentators like Michael Schredl may remain skeptical as they object to my calling dream visions "hallucinations" and dream false beliefs "delusions" on the grounds that those are pathological events which can also occur in waking. I regard these admittedly interpretive definitions as conceptually valid and consensually well established. Dreaming is sufficiently like psychosis as to warrant this comparison. Dominic Ffytche gives us a useful discussion of variations on the theme of hallucination; for Ffytche, dream visions are not properly called hallucinations. He prefers to reserve that term for the false perceptions of waking; I am satisfied that he agrees that my study of hallucinosis in dreaming is heuristically valuable. Ffytche asserts that dreaming hallucinosis and wake state hallucinations must use the same brain substrates. That's more than enough consensus for me.

I specifically challenge Michael Schredl to document his claim that all psychotic patients realize that their symptoms are not real. Most of my psychotic patients were convinced of the reality of their subjective experiences. Indeed, I spent many futile years trying, often unsuccessfully, to instill insight about psychotic symptoms via psychotherapy and potent antipsychotic medication. Like psychiatrist commentators Silvio Scarone and Armando D'Agostino, most scientists accept the dream-psychosis analogy. This is an important assumption of protoconsciousness theory as

I will further emphasize below. The physiology of REM sleep is thus germane to understanding the mechanism and function of both psychosis and normal dreaming.

42.3 What Is the Relationship of Dreaming to Protoconsciousness Theory?

Protoconsciousness is defined as a preverbal mental state that is correlated with fetal and early infantile REM-like brain activation but not with repressed infantile wishes as Freud maintained (Freud 2000). This concept is introduced to carry forward Gerald Edelman's definition of primary consciousness, a state of awareness of protoself, sensorimotor integration, and emotion that might be present in non-human mammals (Edelman 1992). Adult dreaming adds degrees of secondary consciousness to the experiential mix but these secondary features are nonetheless weak in dreaming compared to waking. Hence, for example, *pace* Antrobus and Schredl, abstract reasoning is markedly impaired in dream consciousness. We are thereby given a look at early, preverbal, and primitive protoconsciousness when we dream.

From a developmental perspective, I suggest that protoconsciousness and dreaming both precede and pave the way for waking consciousness in which secondary features such as sensory extero-perception, voluntary movement, language and other abstract forms of cognition are more and more prominently present. Their absence from dreams is suggested and recognized as a scientific opportunity by the Italian linguist, Andrea Moro, who says he never does math in his dreams. Neither do I but I don't do math in waking either. I have examined over 1,000 laboratory REM sleep dream reports and never encountered a single case of dream mathematics. The Moro commentary attests to our reluctance to see a scientific opportunity when it is staring us in the face. It is gratifying to suppose that a linguistics of dreaming might emerge from our dialogue. Moro more than makes up for the absence of the invited but declined commentaries of Steven Pinker and Noam Chomsky.

Many of the invited commentators who did respond are comfortable with these ideas (chief, and most welcome among them, are neurologists Allen Braun and Antonio Damasio along with philosophers Thomas Metzinger and Jennifer Windt). In his commentary, psychologist Michael Eysenck supports many of the key tenets of my theory and seems to be comfortable with my strategy of brain-mind correlation in dream consciousness science. Since I was inspired early in my career by the example of his father, Hans Eysenck, I was pleased to meet Michael at the University of Roehampton where both he and Dominic Ffytche discussed the themes of these lectures in 2010.

Ludwig Huber's discussion of behavioristic studies of non-human animal consciousness is particularly germane to the linguistic issues discussed by Andrea Moro. I am reminded of Ernest Nagel's provocative question, "What is it like to be a bat?" Indeed, bats are non-human mammals which evince REM sleep. Do bats dream? We can only guess what bats subjectively experience. That is equally true of protoconscious subjective experience in humans. I submit, however, that the guesswork is justified if, but only if, the guesses turn into testable hypotheses and they do, as I will attempt to show in my new book, *Psychodynamic Neurology* (Hobson 2014).

42.4 What Other Cognitive Properties Are Altered in Dreaming?

The intensification of negative emotion is one cognitive function of dreaming which commentators Joseph LeDoux and Edward Pace-Schott discuss. LeDoux is understandably focused upon wake-state anxiety since his work has revealed so much about the neurobiology of behaviorally conditioned fear. Sadly, LeDoux says nothing about what stimulus inputs might trigger the often nightmarishly strong dream anxiety. Surely they are not the external inputs that LeDoux employs in his experiments. The recently published fMRI study of Charles Hong (Hong et al. 2008) reveals that many thalamic structures are physically activated together with the REM sleep eye movements. Hong interprets these findings in terms of the well-known PGO waves that are found in cats and now clearly exist in humans, too. It makes good sense to suppose that these internal stimuli trigger the sensations of dreams and that they may trigger emotion like the fear anxiety so well documented by LeDoux.

Edward Pace-Schott, who was my student, has taken the neurobiological analysis of dreaming to a much more elaborate theoretical stage than I ever imagined possible. According to Pace-Schott, the amygdala is only one of many brain nuclei involved in processing dream emotion (for details, see Fig. 3.1, which was designed by Ed Pace-Schott). This figure indicates that the amygdala may be a necessary but insufficient region for the mediation and integration of dream emotion. This idea, in turn, cautions against overconfidence in our all too narrow brain localization efforts.

Since no commentator addressed my suggestion that the profile of dream emotion suggests Darwinian survival functions for anxiety-fear, happiness-elation, and anger, I can only assume that my esteemed peers were too polite to contradict me or that they are not particularly interested in this part of protoconsciousness theory. I had hoped that psychologists Antrobus and Schredl would weigh in on this issue since they so stubbornly refuse to accept any scientifically documented difference between wake and dream consciousness. Perhaps their conservative resistance is useful. Allen Braun did ask why social emotions like shame, guilt, and sadness were so underrepresented in dreams (which typically represent socialization experience). I cannot explain this paradox except to suggest that the emotion profile of dreams reveals innate predictions while the social content reveals the current environment. Thus it is possible that there is a dissociation between genetic and environmental priority for representation in dreaming. It is also true that anxiety, approach-avoidance concern, and aggression are, in fact, socially significant emotions.

Michael Schredl claims that a full range of waking thought and emotion is present in his own and his experimental subjects' dreams. I invite him to keep and publish, in a peer reviewed international journal, the analysis of his record of self-observations so that his claims are at least documented and can be vetted by his peers. He may be in for some surprises. When Andrea Kübler and I asked blinded judges to diagnose the gender of dream reporters, we found that, yes, *pace* Schredl, there were statistically significant gender differences that were detectable but that erroneous attributions of gender were nine times more frequent than correct ones! Would Michael Schredl care to publish these results in his on-line International Journal of Dream Research? They clearly indicate that the dreams of men and women are more the same than different. I predict that wake state controls, not yet available, will also show that men and women have a more similar than different subjective experience. My personal preference is for "*Vive la difference*"; even if the difference is small, it is fun to play with!

These issues should be addressed in those future studies of dream cognition and emotion, which take the altered physiology into account. It is no longer acceptable either to promote Cartesian dualism by assuming that the brain and the mind are separate entities which are impossible to study together nor is it any longer cogent for psychologists to ignore physiology or attempt to explain it away. It is my contention that the physiological difference between REM and wake are so strong that any failure of psychologists to find state dependent differences in consciousness must be the result of scientific failure and/or deep bias, or both.

The apparent lapses in attention in dreaming were discussed by Michael Posner, an integrative experimental psychologist who has pioneered the study of attention in waking human subjects. Posner points out that the control function of attention is lost in dreaming but the awareness function is fully present. Attention in dreaming cries out for attention (sic) using Posner's tests and concepts (Posner and Rothbart 2007) and armed with the affirmative probes for dreams that physicist David Kahn and I have developed (Kahn and Hobson 1994, 2005). Psychiatrist David Spiegel's consideration of the hypnosis-dreaming comparison might take a page from Posner's book: yes, attention is intense in hypnosis but, like dreaming, it is the awareness component of attention that is heightened while the control component is lost unless the dreamer becomes "lucid."

42.5 Who Still Defends Psychoanalytic Dream Theory?

Dan Dennett sees me as quixotic in pounding on an open door of the office that used to be Freud's. Let me point out that the Freud myth persists often covertly but sometimes overtly.

Mark Solms, for example, refuses, as usual, to answer my explicit questions to him about Freud's theory of wish fulfillment as dream instigator and disguise censorship as the reason for dream bizarreness. I feel strongly that these fundamental tenets of Freud's dream theory are erroneous and now are supplanted by the more reasonable and more solid alternatives proffered by brain research. I can only suppose that Solms agrees with me on these points but is politically restrained from saying so in print. I am flattered that he views my work as supporting Freud on other points but I regard this flattery as both disingenuous and insensitive especially when it concerns the similarities and differences between my borrowing from Edelman the primary-secondary definition of consciousness and Freud's two process model. Freud's two process model was designed to foster the notion of dreams as a manifestation of a dynamically repressed unconscious. This misunderstanding may be a good debating strategy but it is not good scientific argumentation. Freud's theory is flatly incorrect on fundamental points. Repressed infantile wishes have nothing to do with dream instigation and dream bizarreness does not result from a defensive need to keep those wishes from invading consciousness. Such glaring errors need to be acknowledged and weighed in the balance, especially as so many of Freud's other ideas derive from them.

Solms' own neuropsychological study of the loss of dreaming (or, more precisely, the loss of the reporting of dreaming) in stroke patients with lesions of the forebrain is important and is interesting. However, Solms' neuropsychological findings (even if confirmed in the sleep lab) in no way counter or question the interaction of brainstem neuromodulation with the forebrain. My view of the modulated forebrain includes the regions thought to be essential for dreaming by Solms.

As for Mark Solms' suggestion that our differences on these issues will be settled "once and for all" (and presumably in Freud's favor) if he demonstrates that damage to forebrain regions (which may well be essential to dreaming) result in poor sleep (thus confirming, according to Solms, (Solms 1997) the logically lame "guardian of sleep" functional hypothesis of Freud). Sleep is guaranteed physiologically, not psychologically; the brain activation of sleep is physiologically engendered by mechanisms that simultaneously and actively block external sensory input. Sleep is thus preserved by intrinsic brain stem processes that are unlikely to be affected by subjective experience or the lack of it.

Regarding consciousness, I argue that these proposed experiments are conceptually flawed in that there is nothing that needs guardianship if infantile wishes are not dream instigators and if dream bizarreness is not defensive. Finally, if not controlled properly and conducted in a sleep lab, Solms' illogical experiments will be without empirical scientific merit. I maintain that Solms' results should be published in a peer-reviewed journal, not a book or a discursive review paper. Peer reviewed journals do not include those of which an author is editor.

I never thought, nor did I ever claim, that dreaming could occur in any decerebrate animal, be it man or beast. In other words, even if I accept Solms' stroke findings, I do not see how they confirm Freud. I have always held that a healthy forebrain was essential to dreaming. Steven Laureys and Vanessa Charland-

Verville emphasize this point in their commentary about the devastating effects of brain stem damage to waking cognition and consciousness. In sum, if Mark Solms regards the disguise-censorship tenet of psychoanalytic dream theory to be untenable, I wonder what is left of Freud's dream model to defend?

There are still lots of closet Freudians around and they turn up in surprising places. One surprising place was the Arizona Conference on Consciousness where two thirds of the audience at my 2006 debate with Mark Solms agreed with him that psychoanalytic theory should be retained. This was surprising to me because the audience was comprised of not only psychologists and psychiatrists (who might have vested interest in Freudianism) but also neurobiologists and even physicists (whom one might expect to have known better).

The main complaint about my position in Arizona was that activation-synthesis did not really replace Freud's theory in its many details. I have recently written an extensive critique of psychoanalytic theory that goes far beyond the obsolete dream theory of Freud to question the concept of defense and related psychological assumptions. A summary of that book, entitled Ego Ergo Sum (Hobson 2013), is available and I will happily dialogue with any respectful scholar about my claims.

I had already persuaded the vast majority of my peers that the dream theory was deficient and argued that since the dream theory is the foundation stone of psychoanalysis, the whole edifice was fatally undermined by the demonstrable defects of the dream theory. I still am sure that this is true and have since both fleshed out protoconsciousness theory and explored its clinical implications. Claus Vögele anticipates these conjectures in his commentary written from the perspective of clinical psychology. Vögele is an expert in the science of anxiety and, like many practitioners of psychotherapy, has given up speculation about early-life trauma as conceived by Freud in favor of an environmentally sensitive version of protoconciousness theory that fits his research commitment to psychophysiology.

Maybe the world will have to wait until card-carrying psychoanalysts and their many followers are dead before we see a younger breed of psychological thinkers rise from the ashes.

References

Edelman, G. M. (1992). *Bright air, brilliant fire: On the matter of the mind.* New York: Basic Books.

Freud, S. (2000). *The interpretation of dreams* (Translated from German and edited by J. Strachey). New York: Basic Books.

Hobson, J. A. (2013). Ego ergo sum: Toward a psychodynamic neurology. *Contemporary Psychoanalysis, 49*(2), 142–164.

Hobson, J. A. (2014). *Psychodynamic neurology: Dreams, consciousness, and virtual reality.* New York: CRC Press.

Hobson, J. A., Pace-Schott, E. F., & Stickgold, R. (2000). Dreaming and the brain: Toward a cognitive neuroscience of conscious states. *Behavioral and Brain Sciences, 23*(6), 793–842.

Hong, C. C., Harris, J. C., Pearlson, G. D., Kim, J. S., Calhoun, V. D., Fallon, J. H., Golay, X., Gillen, J. S., Simmonds, D. J., van Zijl, P. C., Zee, D. S., & Pekar, J. J. (2008). fMRI evidence for multisensory recruitment associated with rapid eye movements during sleep. *Human Brain Mapping, 30*(5), 1705–1722.

Kahn, D., & Hobson, J. A. (1994). Self-organization theory of dreaming. *Dreaming, 3*, 151–178.

Kahn, D., & Hobson, J. A. (2005). State-dependent thinking: A comparison of waking and dreaming thought. *Consciousness and Cognition, 14*(3), 429–438.

Posner, M. I., & Rothbart, M. K. (2007). Research on attention networks as a model for the integration of psychological science. *Annual Review of Psychology, 58*, 1–23.

Solms, M. (1997). *The neuropsychology of dreams: A clinico-anatomical study*. Hillsdale: Erlbaum.

Chapter 43
Lecture II: Physiology

J. Allan Hobson

Most of the initial resistance to reciprocal interaction has died away in the face of substantial confirmation of its claims and its adoption by many clinical formulations and general models of sleep. For that reason comment was not invited from the basic sleep research community. My worthy physiologist colleagues tend to consider my interest in dreaming as ill advised as did the NIMH, which nonetheless supported the physiological research of my laboratory for 27 years.

It is gratifying to read, in the commentary of basic neuroscientist Rodolfo Llinás, his opinion that differences between waking and dreaming consciousness might well correlate with the decrement in neuromodulation as well as the blockade of sensory input known to characterize REM sleep. I was disappointed that other distinguished basic physiologists declined my invitation to comment; for example, Eric Kandel, a psychiatrist and fellow-trainee at Harvard, turned me down on the grounds that sleep research was beyond his ken. On its face, this reason seems valid enough but it was unsatisfying to me in view of Kandel's elegant demonstration that the failure of the aplysia ganglion to learn the gill retraction reflex was due to the fact that the L 15 cell (containing serotonin) was not in the loop (Kandel 2004). This fact struck me as begging comparison with dream amnesia and the REM sleep arrest of serotonin neurons firing in the cat. It seems quite likely that serotonin demodulation also occurs in human REM and I therefore suggested that it might be a physiological mechanism of dream amnesia. Learning and memory are strongly shared interests of both invertebrate and vertebrate oriented neurobiologists.

J.A. Hobson (✉)
Department of Psychiatry, Harvard Medical School, Boston, MA, USA
e-mail: allan_hobson@hms.harvard.edu

N. Tranquillo (ed.), *Dream Consciousness*, Vienna Circle Institute Library 3,
DOI 10.1007/978-3-319-07296-8_43,

43.1 What More Can We Expect to Learn About Dreaming by Studying the Physiology of REM Sleep?

A great deal, if commentators Martin Dresler and Allen Braun are correct, and I think that they are. PET scan results have already given the science of human dreaming a strong and direct boost. Three groups (including that of Allen Braun himself) have shown that in man, as in the cat, the pontine tegmentum is activated above the level of quiet waking (Braun et al. 1997; Maquet et al. 1996; Nofzinger et al. 1997). So are several forebrain structures including the parietal operculum and deep frontal white matter, which the Solms stroke studies suggested might be critical for dreaming. Joseph LeDoux must be pleased to see the amygdala light up in REM sleep, perhaps helping us to explain dream fear.

As Martin Dresler points out in his commentary, a great promise for the future of dream science resides in the numerous advantages over PET of fMRI brain scanning technology. fMRI has already been used to analyze lucid dreaming (which I treat as a separate topic following this one). Dresler and his colleagues at the Max Planck Institute in Munich hope to avail themselves of the real time, on-line analysis of human brain activity as subjects wake and perform cognitive tasks. The subjects then try to sleep in the scanner while their brain, eye, and muscle activity is monitored continuously. Such subjects can be awakened in REM and reports of mental activity elicited. These experiments are expensive and difficult to perform but their promise is substantial.

43.2 How Can Lucid Dreaming and Its Study Advance the Protoconsciousness Agenda?

Lucid dreaming is a hybrid state bred of an unlikely but clear admixture of primary and secondary consciousness features. As such it constitutes proof that waking and dreaming can co-exist and be teased out to show that different states of different regions of the same brain underlie waking and dreaming. The idea that waking and dreaming are admixtures of two distinct but cooperative brain processes is thus validated by lucid dreaming. Ursula Voss, who has pioneered this new paradigm, offers a commentary that is based upon her important data. Again, the phenomenological differences between waking, lucid, and non-lucid dreaming are evidence that we can now speak of three distinct (if overlapping) brain-mind states.

With my enthusiastic support and guidance, Voss has applied the quantitative EEG (qEEG) method to establish the following facts: when subjects become lucid they increase their self-awareness as they increase the 40 Hz power of their frontal EEG (Voss et al. 2009). This confirms the commentary of Wolf Singer who has written the foreword to this volume and has hypothesized that waking consciousness results from high frequency synchronization of the forebrain. Dream lucidity could be a by-product of increased 40 Hz activation in REM. Voss has recently published the scale by which lucidity as a subjective experience can be measured (Voss et al. 2012). She has also succeeded in raising the intensity of 40 Hz EEG power and the intensity of lucidity in REM sleep dreaming by stimulating the frontal scalp electrically. I must admit that I advised against this experiment and, to my delight, have been proved wrong (Voss et al. 2014).

43.3 Sleep and Dream Science Enriches Our Understanding of Learning

Suzanne Diekelmann and Jan Born provide a lively and trenchant discussion of the evidence concerning sleep's role in the consolidation of the learning experiences of waking. I commend their commentary with which I mainly agree. The evidence is compelling that we do improve both our explicit and implicit learning while we sleep. Diekelmann and Born do not deal with the paradox of the learning enhancement by states which are, themselves, so poorly remembered. I wonder if this means that we are not meant to be aware of many basic brain housekeeping functions including learning, body temperature control, rest and dreaming itself. This paradox has prompted Owen Flanagan to suggest that dreaming is a spandrel of sleep (spandrel meaning a functionless decoration). José Cantero and Mercedes Atienza, who worked with me and Robert Stickgold on memory consolidation before setting up their own lab in Seville, Spain, are also silent on this paradoxical question.

By far the most analytical approach to the sleep-learning story is commented upon by Giulio Tononi, the psychiatrist-physiologist now working in Madison, Wisconsin. Giulio Tononi has performed very sophisticated studies which indicate that highly localized regions of the cortex process the previous day's information while mammals sleep. This could long ago have been inferred from dreams were we not so addicted to their prophetic interpretation. Tononi's experiments do not rule out prophecy any more than my own neurophysiology does, but our much more modest claims may end up being more bold than the promissory trumpeting of dream symbolists. Now I dream of a Tononi-like champion and investigator of protoconsciousness!

43.4 Dreams, Vision, and PGO Waves

Psychologist commentators John Antrobus and Don Kuiken take a very dim view of the relationship of eye movements themselves (and the PGO waves with which they are associated in cats and humans) to the subjective experience of dreaming. They prefer to side with veterinarian-physiologist Adrian Morrison's commentary asserting that PGO waves signify the activation of a startle network having little or nothing specific to do with dreaming. I agree that, to date, the specificity of non-REM sleep phasic physiological activity correlation with the microscopic structure of dream cognition has not yet been established. In this connection, I would have welcomed a disclaimer from William Dement, who first championed the eye movement-hallucinatory gaze direction hypothesis of dreaming in the 1960s. I hoped that Dement, a generous and modest pioneer of REM-dream science, might have echoed my "not yet" caution and advised an open mind and scientific persistence. He might have warned that the absence of proof is not the proof of absence.

In the absence of proof, I submit that there can be little question that PGO waves encode upcoming eye movement direction in waking, in startle, and in REM sleep. Protoconsciousness theory thus asserts only that PGO waves are internal signals from the motor to the sensory side of the nervous system and that the excitability of the PGO generator system of the pons is greater in startle than in quiet waking and greater still in the complete absence of external sensory data during REM.

Adrian Morrison reviews his startle network theory in his commentary. In response, I argue that it is rigorously established that PGO waves have a functional role in all three states: waking, startle response, and dreaming. Furthermore, the difference in PGO wave activity between those three states is profound and potentially informative. A reader seeking details regarding this argument should consult my recent paper (Hobson and Friston 2012) on the subject, co-authored with psychiatrist-mathematician, Karl Friston, whose commentary showcases the Helmholtzian free energy paradigm (see my summary of that arcane but prescient theory in Chapter 44).

The protoconsciousness theory likens dreams to blindsight (briefly commented upon by the originator of that paradigm, Larry Weiskrantz). Dreams are a variation on the theme of blindsight because, as Leonardo da Vinci once declared, dream vision can be as sharp as it ever is in waking despite the absence of external visual sensory stimuli. It would surprise me greatly if the brain made no use of this informational specificity and if this specificity was not somehow related to the visual aspect of dreams. In any case, I theorize that visual system activation in sleep is a kind of warm-up for vision in subsequent waking. That aspect of protoconsciousness theory apparently excited none of my invited commentators.

References

Braun, A. R., et al. (1997). Regional cerebral blood flow throughout the sleep–wake cycle. *Brain, 120*, 1173–1197.

Hobson, J. A., & Friston, K. J. (2012). Waking and dreaming consciousness: Neurobiological and functional considerations. *Progress in Neurobiology, 98*, 82–98.

Kandel, E. R. (2004). The molecular biology of memory storage: A dialog between genes and synapses. *Bioscience Reports, 24*(4–5), 475–522.

Maquet, P., Péters, J., Aerts, J., Delfiore, G., Degueldre, C., Luxen, A., & Franck, G. (1996). Functional neuroanatomy of human rapid-eye-movement sleep and dreaming. *Nature, 383* (6596), 163–166.

Nofzinger, E. A., Mintun, M. A., Wiseman, M. B., Kupfer, D. J., & Moore, R. Y. (1997). Forebrain activation in REM sleep: An FDG PET study. *Brain Research, 770*(1–2), 192–201.
Voss, U., Holzmann, R., Tuin, I., & Hobson, J. A. (2009). Lucid dreaming: A state of consciousness with features of both waking and non-lucid dreaming. *Sleep, 32*(9), 1191–2000.
Voss, U., Frenzel, C., Koppehele-Gossel, J., & Hobson, J. A. (2012). Lucid dreaming: An age-dependent brain dissociation. *Journal of Sleep Research, 21*(6), 634–642.
Voss, U., Holzmann, R., Hobson, J. A., Paulus, W., Koppehele-Gossel, J., Klimke, A., & Nitsche, M. A. (2014). Induction of self awareness in dreams through frontal low current stimulation of gamma activity. *Nature Neuroscience, 17*(6), 810–812.

Chapter 44
Lecture III: Philosophy

J. Allan Hobson

William James might have been pleased, were he alive, to see so many of his philosophical descendents so vigorously involved in a discussion of dream consciousness – a topic he treated surprisingly little. Taken all in all, I think a good case can be made for a return to the pre-Jamesian spirit of "natural philosophy," where that term denotes the application of critical thought to the science of phenomena. One does not have to be a professional philosopher to be a natural one and professional philosophers need neither eschew nor ignore the nature that attracts experimental scientists.

Many of the commentaries on my Dream Consciousness lectures reveal a gratifying commitment to the integration sanctified by William James. Mind-brain integration involves an encouraging degree of boundary crossing. Some of the best examples of fearless boundary crossing are given by commentators Allen Braun and Antonio Damasio, both neurologists who are not afraid of psychiatry and philosophers Thomas Metzinger and Jennifer Windt who immerse themselves in ego science – and even collect experimental data about it. Thanks to fearless peers like these, specialist fields are united in what can properly be called natural philosophy. Metzinger and Windt share an interest in the "phenomenological self" and are thus sensitive to my protoconsciousness model's self-as-agent concept. I have long believed that a more critical look at self-representation in dreams was in order. Perhaps the time for such a study is now.

Contributions to this discussion might have included philosopher Owen Flanagan (who was very enthusiastically engaged in the discussions in Vienna) and the Dean of Neurophilosophy herself, Patricia Churchland. I am confident that this volume will receive their blessings in the sad absence of their written commentary. Other more understandably absent friends include John Locke and Immanuel Kant, whose deaf-because-dead ears might be imagined to burn with

J.A. Hobson (✉)
Department of Psychiatry, Harvard Medical School, Boston, MA, USA
e-mail: allan_hobson@hms.harvard.edu

N. Tranquillo (ed.), *Dream Consciousness*, Vienna Circle Institute Library 3,
DOI 10.1007/978-3-319-07296-8_44,

disdain and pleasure respectively, as we name them. The dreaming brain says that it has *a priori* knowledge and that it is anything but a *tabula rasa*. It would have been nice to have contributions from Karl Popper (Do you still think that psychoanalytic theory is not scientifically testable?) and John Eccles (Do you still believe that dreaming results from a separation of mind from body?). Would Sigmund Freud have accepted an invitation to answer the question? (Is your 1895 Project for a Scientific Psychology now within our grasp?) Such thought experiments are in the tradition of philosophy.

As usual, Dan Dennett gives us the best of philosophical thought itself with his dazzling display of alternative speculations about the mechanism and function of dreaming. He is quite right to mention a myriad of other ways of looking at these data but I have both moral and intellectual discomfort with some of his suggestions: first, I suppose that REM deprivation of infants and fetuses is both impossible and morally reprehensible so that direct test of the protoconsciousness hypothesis is not, unfortunately, available. I argue for the plausibility and explanatory power of my theory. My second concern is with the plausibility of Dennett's own suggestion that the persistence of REM and dreaming in adult life is a functionless appendage (like the appendix in animals that no longer munch grass). The demonstrated utility of the REM sleep enhancement of procedural learning clinches this point empirically. We may no longer munch grass but we must husband energy and body temperature.

These are as strong data as any that Charles Darwin could adduce in direct support of his theory of evolution, a theory that Dennett rightfully admires. I have similar reservations about Flanagan's charming "spandrels of sleep" theory of dreams. While a part of me is sympathetic with the implied epiphenomenalism, I suffer nagging doubts about dismissing the mind's causality, a topic which I treat separately below.

Dennett stops short of engaging in experimentalism himself and wisely sidesteps the blandishment of endorsement and the straightjacket of yes and no answers to questions. The Dennett commentary shows that philosophy is a very free country and I am happy for that.

More narrowly construed, philosophy comprises logic and epistemology. Taking the second component first we can take satisfaction that there is an increasing reliance by epistemologists on experimental evidence (John Searle is a good example) and there is widespread recognition that philosophy of mind cannot advance without neurobiology. I have already asserted that psychology cannot afford, any longer, to go it alone. In the following section I earmark my own scientific logic and strategy for closer attention by philosophers and then close with a discussion of the science of function, an important concern of any natural philosophy.

44.1 The Formal Mental Status Paradigm

It was a great disappointment to me that so few commentators either praised or criticized the formal mental status approach that I have used in my psychophysiological studies of dream consciousness. Thomas Metzinger and Jennifer Windt seem to understand why I take the formal rather than the content-based approach. Perhaps I should have directed Michael Schredl's critical attention to the formal strategy so widely applied by me in the dream psychology data summarized in Table 1.1. The precedence of form over content in humanistic as well as scientific studies is emphasized by the art historian commentator, Hellmut Wohl, who made this point in his own work long before he collaborated with me. Nor was any comment made by philosophers about my invocation of brain-mind isomorphism. I freely admit that my opening lecture is anecdotal in its formal analysis of a single dream report. It was designed to illustrate rather than convince. I hope that Daniel Dennett and Antti Revonsuo will let me know whether they accept my arguments in favor of the formal treatment of dream reports.

44.2 The Sleep Lab: A Cloudy Lens?

I encourage the development of home-based approaches to dream science because I regard the home as a more commodious and natural place to sleep and because, having slept in the lab as an experimental subject, I know at first hand how difficult it is to wake up on demand and how suggestible are the sleep-fogged minds of even young experimental subjects (and an additional problem is that most experimental subjects *are* young). This means that even the half-baked conclusions of the first half-century of dream science are age biased and heavily so. My reasons for optimism about field work in dream science are given in Fig. 2.3 and Table 2.1. No one commented on this serious issue so I will now state clearly that I believe a new initiative of this kind to be essential. The good news is that such an initiative lends itself to college and even high school student and faculty participation in dream science.

Home-based studies are admittedly invasive but they are inexpensive and interesting to carry out. With funding agencies channeling their resources into high tech studies of everything (including REM) there is little hope for support in a field that has been officially considered to be outside the bounds of science. This is unfortunate because it is now possible to imagine a reconstructed, brain-based approach to psychodynamics making itself useful and self-instructive here. The commentaries of Braun and Damasio make this point indirectly but clearly.

Modern psychotherapists (and there are thousands of them) can join this scientific endeavor by training themselves and their patients to get on board. I hope to live long enough to make my views known to young clinicians. Meanwhile I commend the book, *Night*, by the London-based author, Al Alvarez who slept in a sleep lab and describes his own confusion when experimentally awakened (Alvarez 1995).

44.3 Energy and Information Regulation

The old wives' advice about getting a good night of sleep (implying a much needed rest) is pop psychology in search of science. This search was long ago undertaken by the great German scientist, Hermann von Helmholtz. Helmholtz beat Freud to the punch, even in terms of dream theory, by enunciating the doctrine of free energy in the mid-nineteenth century. Part of that theory, the prediction of the sensory effects of movement, is relevant to protoconscious dream theory as Helmholtz makes clear in ten trenchant pages in his masterful book entitled *Physiological Optics* (von Helmholtz 2000).

When Jack Nelson studied the lateral pontine tegmental cluster of PGO burst cells, I recognized the relevance of those findings to Helmholtz' idea of feed-forward excitation from the motor to the sensory side of the visual system. This is where Rodolfo Llinás comes in. His commentary does not do his book (Llinás 2001) justice but Karl Friston honors us with a typically sophisticated commentary about the importance of Helmholtz' free energy theory. Friston and I agree that a surprising possibility is that sleep (and especially REM sleep) helps the organism to regulate brain temperature and, at the same time, allow the brain to predict and correct the errors of visual information acquisition (Hobson and Friston 2012).

Who would ever have guessed that dreaming was our subjective experience of so functionally significant a state? Protoconsciousness theory does its best to show why these disparate parts hang together. It is indeed pleasing to see us moving closer to a view of sleep and dreaming that comprises both energy and information. We cannot say what the mind consists of – is it particles or is it waves or neither – but we are no worse off than physicists who grapple with space, time, and the structure of matter. In fact, we might well echo the Elegy Written in a Country Churchyard by Thomas Grey: If ignorance is bliss, tis folly to be wise.

44.4 Creativity

Creativity (including artistic, literary, and scientific originality) is at least as functionally adaptive as our tendency to mystical explanations that arise from our religion-minded brains. Allen Braun tells us in his commentary that improvisational jazz musicians deactivate their frontal cortex when they are in a creative zone in waking just as they do when they are dreaming. At the same time that the dorsomedial nucleus goes dark, the nearby mediodorsal nucleus lights up. This finding, however preliminary, is relevant to two aspects of dream science. One is that dreaming itself is creative (as Hellmut Wohl asserts in his commentary); the other is that aspects of dreaming can be enhanced in waking (just as Ursula Voss' commentary on dream lucidity indicates that aspects of waking can be enhanced in dreaming). This double reciprocity affirms a strong claim of protoconsciousness theory: waking and dreaming are both hybrid states and that their hybridicity can be manipulated to functional advantage. We can condition ourselves to be less fearful in nightmarish sleep and more artful in unimaginative waking. These are both functionally significant possibilities which dispense with alcohol and drugs as they require only autosuggestion.

44.5 Cortical Rewiring, Protoconsciousness, and AIM

A mechanism for achieving plastic changes, including the plastic changes needed to condition learning or to create new forms of cognition, is provided by the science described in the commentaries of Braun, Damasio, Dresler and Tononi. Dream science may thus hold the key to understanding these important functional advantages and give sleep research pride of place in this domain. At first counterintuitive, it now seems natural to suppose that clearing the decks for action may well be facilitated by shutting down inputs and outputs (I) and changing the neuromodulatory chemistry (M) while at the same time activating (A) the cortex and thus simulating but not actually instantiating waking. These considerations are relevant to Revonsuo's and Noreika's suggestions about threat avoidance rehearsal and even inform Crick and Mitchison's suggestion (Crick and Mitchison 1983) that we dream in order to forget. David Hartley would sit up in his grave at the sound of such music.

The bottom line is that radical cerebral housekeeping may be going on while we lie, outwardly unconscious, in our beds at night.

44.6 Is Consciousness Causal?

Whether subjectivity causes anything physical to occur is obviously the first and last question on all of our minds. We might well say, with William James, that "I will believe in free will by way of showing that my will is free." But this famous tenet of pragmatism is not scientifically satisfying because it is, essentially, a religious belief and today we are not as tolerant of religion as James was in 1890. We have greatly improved our scientific capability to the point where we can reopen Freud's 1895 *Project for a Scientific Psychology*, but the consciousness-as-causal is today much more in doubt than it was at end of the 19th century (Freud 1895).

Commentators wisely skirted this puzzle. I am not entirely happy with acceptance of the implications of the negative experimental data of Libet and Wegner (Libet et al. 1983; Wegner 2004). One way out is to suppose that the physical and psychological domains are separate but equal parts of a unified bi-modal system in which case both domains are causal even though there may be time delays in uniting the psychological domain with the physical. It is very difficult to assume that consciousness has no consequential effect on behavior. This is not only a hope that conscious will is not a useless subjective illusion. It is also a scientific concern. If consciousness itself is not causal, then what in the world is it for?

Can dream science help us solve this vexing problem? I think so. A robust truth is that the will, be it illusion or reality, is state dependent. In dreams, our will is dramatically weakened along with other executive ego functions. This is one reason for Owen Flanagan's view of dreams as mere spandrels of sleep. A central aspect of lucid dreams is the restoration of will (or the illusion thereof) suggesting that the frontal lobes mediate voluntary decision making as their pre-motor anatomy might suggest. Believe it or not, volition has yet to be operationally defined and systematically investigated in relation to sleep and dreaming. This astonishing oversight must be redressed, and soon! It certainly can be using the concepts and methods of the protoconsciousness paradigm. The way in which the immaterial mind may exert a causal effect upon the material brain has recently been detailed (Hobson and Friston 2014).

The measurement of conscious will in waking and dreaming might help skeptical critics better understand and appreciate the power of the protoconsciousness paradigm. I know from personal conversation that Antti Revonsuo, one of my most outspoken and respected critics, is sure that dreaming of threat avoidance is essential to the efficacy of fight-or-flight dream rehearsals. I was initially skeptical of Revonsuo's assumption about the causality of subjective experience of dreams and only introduce this anecdote to make a final point: we still have much to learn from the study of dream consciousness. I hope that protoconsciousness theory helps that to happen.

References

Alvarez, A. (1995). *Night*. New York: W.W. Norton and Co.

Crick, F., & Mitchison, G. (1983). The function of dream sleep. *Nature, 304*, 111–114.

Freud, S. (1895). *Project for a scientific psychology*. New York: Standard Edition.

Hobson, J. A., & Friston, K. J. (2012). Waking and dreaming consciousness: Neurobiological and functional considerations. *Progress in Neurobiology, 98*, 82–98.

Hobson, J. A., & Friston, K. J. (2014). Consciousness, dreams, and inference: The cartesian theatre revisited. *Journal of Consciousness Studies, 21*(1–2), 6–32.

Libet, B., Gleason, C. A., Wright, E. W., & Pearl, D. K. (1983). Time of conscious intention to act in relation to onset of cerebral activity (readiness-potential). The unconscious initiation of a freely voluntary act. *Brain, 106*(Pt 3), 623–642.

Llinás, R. R. (2001). *I of the vortex: From neurons to self*. Cambridge, MA: MIT Press.

von Helmholtz, H. L. F. (2000). *Helmholtz's treatise on physiological optics*, 3 Vols. (Translated from the 3rd German ed., J. P. C. Southall, Ed.). London: Thoemmes Press.

Wegner, D. M. (2004). Précis of the illusion of conscious will. *Behavioral and Brain Sciences, 27* (5), 649–659.

Printed by Printforce, the Netherlands